Learning Visual Basic Through Applications

LIMITED WARRANTY AND DISCLAIMER OF LIABILITY

THE CD WHICH ACCOMPANIES THE BOOK MAY BE USED ON A SINGLE PC ONLY. THE LICENSE DOES NOT PERMIT THE USE ON A NETWORK (OF ANY KIND). YOU FURTHER AGREE THAT THIS LICENSE GRANTS PERMISSION TO USE THE PRODUCTS CONTAINED HEREIN, BUT DOES NOT GIVE YOU RIGHT OF OWNERSHIP TO ANY OF THE CONTENT OR PRODUCT CONTAINED ON THIS CD. USE OF THIRD PARTY SOFTWARE CONTAINED ON THIS CD IS LIMITED TO AND SUBJECT TO LICENSING TERMS FOR THE RESPECTIVE PRODUCTS.

CHARLES RIVER MEDIA, INC. ("CRM") AND/OR ANYONE WHO HAS BEEN INVOLVED IN THE WRITING, CREATION OR PRODUCTION OF THE ACCOMPANYING CODE ("THE SOFTWARE") OR THE THIRD PARTY PRODUCTS CONTAINED ON THE CD OR TEXTUAL MATERIAL IN THE BOOK, CANNOT AND DO NOT WARRANT THE PERFORMANCE OR RESULTS THAT MAY BE OBTAINED BY USING THE SOFTWARE OR CONTENTS OF THE BOOK. THE AUTHOR AND PUBLISHER HAVE USED THEIR BEST EFFORTS TO ENSURE THE ACCURACY AND FUNCTIONALITY OF THE TEXTUAL MATERIAL AND PROGRAMS CONTAINED HEREIN; WE HOWEVER, MAKE NO WARRANTY OF ANY KIND, EXPRESS OR IMPLIED, REGARDING THE PERFORMANCE OF THESE PROGRAMS OR CONTENTS. THE SOFTWARE IS SOLD "AS IS " WITHOUT WARRANTY (EXCEPT FOR DEFECTIVE MATERIALS USED IN MANUFACTURING THE DISK OR DUE TO FAULTY WORKMANSHIP);

THE AUTHOR, THE PUBLISHER, DEVELOPERS OF THIRD PARTY SOFTWARE, AND ANYONE INVOLVED IN THE PRODUCTION AND MANUFACTURING OF THIS WORK SHALL NOT BE LIABLE FOR DAMAGES OF ANY KIND ARISING OUT OF THE USE OF(OR THE INABILITY TO USE) THE PROGRAMS, SOURCE CODE, OR TEXTUAL MATERIAL CONTAINED IN THIS PUBLICATION. THIS INCLUDES, BUT IS NOT LIMITED TO, LOSS OF REVENUE OR PROFIT, OR OTHER INCIDENTAL OR CONSEQUENTIAL DAMAGES ARISING OUT OF THE USE OF THE PRODUCT.

THE SOLE REMEDY IN THE EVENT OF A CLAIM OF ANY KIND IS EXPRESSLY LIMITED TO REPLACEMENT OF THE BOOK AND/OR CD-ROM, AND ONLY AT THE DISCRETION OF CRM.

THE USE OF "IMPLIED WARRANTY" AND CERTAIN "EXCLUSIONS" VARY FROM STATE TO STATE, AND MAY NOT APPLY TO THE PURCHASER OF THIS PRODUCT.

Learning Visual Basic Through Applications

Clayton E. Crooks II

CHARLES RIVER MEDIA, INC.
Hingham, Massachusetts

Copyright 2001 by CHARLES RIVER MEDIA, INC.
All rights reserved.

No part of this publication may be reproduced in any way, stored in a retrieval system of any type, or transmitted by any means or media, electronic or mechanical, including, but not limited to, photocopy, recording, or scanning, without *prior permission in writing* from the publisher.

Publisher: David F. Pallai
Production: Electropublishing
Printer: InterCity Press
Cover Design: The Printed Image

CHARLES RIVER MEDIA, INC.
20 Downer Avenue, Suite 3
Hingham, Massachusetts 02043
781-740-0400
781-740-8816 (FAX)
info@charlesriver.com
www.charlesriver.com
This book is printed on acid-free paper.

Clayton E. Crooks II. *Learning Visual Basic Through Applications.*
ISBN: 1-58450-032-8

    Library of Congress Cataloging-in-Publication Data

Crooks, Clayton.
  Learning visual basic through applications / Clayton Crooks II.
    p. cm.
  ISBN 1-58450-032-8
  1. Microsoft Visual BASIC. 2. BASIC (Computer program language)
I. Title.
  QA76.73.B3 C738 2001
  005.26'8--dc21
                          2001003266

All brand names and product names mentioned in this book are trademarks or service marks of their respective companies. Any omission or misuse (of any kind) of service marks or trademarks should not be regarded as intent to infringe on the property of others. The publisher recognizes and respects all marks used by companies, manufacturers, and developers as a means to distinguish their products.

Printed in the United States of America
01 02 7 6 5 4 3 2 First Edition

CHARLES RIVER MEDIA titles are available for site license or bulk purchase by institutions, user groups, corporations, etc. For additional information, please contact the Special Sales Department at 781-740-0400.

Requests for replacement of a defective CD must be accompanied by the original disc, your mailing address, telephone number, date of purchase and purchase price. Please state the nature of the problem, and send the information to CHARLES RIVER MEDIA, INC., 20 Downer Avenue, Suite 3, Hingham, Massachusetts 02043. CRM's sole obligation to the purchaser is to replace the disc, based on defective materials or faulty workmanship, but not on the operation or functionality of the product.

Dedication

This book is dedicated to my wife Amy. I'm fortunate to have the opportunity to spend every day of my life with you. Thank you for your love and support.

Contents

1. Introduction to Visual Basic — 1

   The Toolbars — 4
      Standard Toolbar — 4
      Individual Toolbars — 5
      The Form Editor Toolbar — 5
   The Windows — 6
      The Toolbox — 6
      Form Window — 9
      Code Window — 9
      Project Explorer — 10
      Properties Window — 11
      Form Layout Window — 11
      Changes in Visual Basic.NET — 12
      Upgrading Version 6 Projects to Visual Basic.NET — 12
      Working with Both Visual Basic 6.0 and Visual Basic.NET — 13
   A Quick Project — 14
      The User Interface — 14
   The Code Explanation — 17
   Running the Program — 18
   Complete Code Listing — 19
   Chapter Review — 19

2. Developing a Visual Basic Multimedia Player — 21

- Project Overview — 22
  - *The User Interface* — 22
- Programming the CD Player — 28
- CD Form Load — 29
  - *Audio CD Commands* — 29
- The Multimedia Player — 30
- Ending the Application — 32
  - *Query Unload* — 32
  - *Unloading the Form* — 34
- Testing the Application — 34
  - *Complete Code Listing* — 36
- Chapter Overiew — 38

3. Creating an MP3 Player — 39

- Project Overview — 40
- The Project — 40
- Setting Things Up — 44
- Playing the MP3 Files — 46
  - *Reading the Tag Information* — 47
- The Rest of the Functions — 50
  - *Tidying Up* — 52
- Complete Code Listing — 53
- Chapter Review — 57

4. Project Overview — 59

- Using the Windows 32 API — 59
  - *API and DLL Advantages* — 61
- The Fundamentals of API Calls — 61
- Project Overview — 62
- The API Calls — 63
  - *Creating Rectangles* — 63
  - *Creating Ellipses* — 63

Creating Rounded Rectangles	63
Create Polygons or Custom Shapes	63
Combining Regions	64
Initializing the Window	64
Setting Up the Form	64
Some Declarations	66
Changes to the Form_Load Event	67
Moving the Form	70
Finishing the Project	70
Complete Code Listing	71
Chapter Review	79

5. Creating a VBPong Game — 81

The Project	82
The GUI	82
The Code	85
The Form_Load Event	85
Paddle Movement	86
The Timer	87
Ball Movement	87
Scoring	90
Complete Code Listing	91
Chapter Review	95

6. A 3D Model Viewer — 97

Project Overview	98
What is Direct3D?	98
Advantages of Direct3D	98
The SDK	98
Direct3D Coordinate System	99
Direct3D Modes	100
3D Vocabulary	100
Project Framework	101

	Declarations	101
	Form_Load Event	102
	Initialize DirectDraw and Direct3D	103
	Creating the Scene	105
	The Main Loop	106
	Closing the Application	107
	Complete Code Listing	108
	Chapter Review	112
7.	**Visual Basic Screen Saver**	**113**
	Project Overview	114
	Beginning the Project	114
	The Label Control Properties	115
	The Timer Properties	116
	Program Declarations	116
	Form_Load	116
	Generate Random Colors	118
	Drawing Gradients	119
	The Second Timer	120
	Ending the Screen Saver	120
	Testing the Project in the IDE	121
	Compiling and Using the Screen Saver	122
	Complete Code Listing	123
	Chapter Review	126
8.	**Design a Word Processor**	**127**
	The Project	127
	Overview	127
	Introducing the Rich TextBox and Common Dialog Controls	128
	Designing a Menu	129
	Adding Some Code	132
	File Menu	132
	Common Dialog Control	133

		Printing	134
		Cut, Copy and Paste	135
		The Font Menu	136
		The Final Steps	138
	Complete Code Listing		138
	Chapter Review		142

9. FTP Program — 143

- **FTP Overview** — 144
- **Beginning the Project** — 144
  - Basics of the INET Control — 144
  - Creating the GUI — 145
- **Writing Some Code** — 146
  - Variable Declarations — 146
  - Connecting to an FTP Server — 147
  - State Changed Event — 148
  - Retrieving Data — 150
  - File Download — 151
  - Finishing Touches — 151
- **Complete Code Listing** — 152
- **Chapter Review** — 155

10. MDI Web Browser — 157

- **MDI Overview** — 158
- **Beginning the Project** — 159
  - MDI Parent and Child Forms — 159
  - Creating the Menus — 160
- **Some Code** — 161
  - Dynamic Creation of Child Forms — 161
  - Child Form Menus — 163
  - The Internet Control — 164
- **Programming the Browser Control** — 164
  - The GUI — 164

Creating the Browser	166
Web Browser Methods and Properties	166
Resizing the Web Browser Control	166
Form Load	167
Complete Code Listing	168
Form1	168
MDIForm1	169
Chapter Review	169

11. Creating a Chat Program — 171

Winstock Control	172
Overview	172
Properties	172
Methods	172
Creating the Forms	173
Server Form	173
The Client Form	173
Startup Form	174
Writing Code for the Client and Server	176
Server	176
Client	177
Testing the Program	177
Complete Code Listing	179
frmClient	179
frmServer	179
frmStart	180
Chapter Review	180

12. Animated Desktop Assistant — 181

Agent Overview	182
Installing Microsoft Agent	182
Creating the Desktop Assistant	183
Placing the Control on Your Form	183
Writing Some Code	184

	Declaring an Agent Control Reference at Runtime	184
	Adding Speech and Animation	185
	Animation	185
	Speech Recognition	187
	Adding Commands	187
	The GUI	189
	Executing the Commands	190
	Complete Code Listing	193
	Chapter Review	194
13.	**Creating an Internet Time Retrieving ActiveX Control**	**195**
	ActiveX Project	196
	The GUI	196
	Developing the Control	197
	Properties	199
	Testing the Control	200
	Creating Another Project	200
	Complete Code Listing	201
	UserControl	201
	Standard EXE Project	202
	Chapter Review	202
14.	**Visual Basic Paint Program**	**203**
	The Paint Project	204
	GUI	204
	Toolbars	205
	Pop-Up Menu	207
	Writing the Code	207
	API and Variable Declarations	207
	Form Load	208
	Menu Events	209
	Changing Colors	210
	Width, Height, and Style	211

Moving the Toolbars	212
The Commands	213
Undo	214
Drawing to the PictureBox	214
MouseDown Event	214
MouseMove and MouseUp Events	215
Complete Code Listing	219
Chapter Review	227

15. Screen Capture — 229

The Program	229
Print Screen Key	230
Retrieve Image from Clipboard in Visual Basic	230
Keybd_Event API	233
BitBit	235
Complete Code Listing	240
frmKeybdEvent	241
frmMain	241
frmPrintScrn	242
Chapter Review	242

16. Graphing Calculator — 243

The Project	243
Creating a GUI	243
Finishing the GUI	248
Writing the Code	251
The Command Button Array	251
Negative or Positive	253
Trig Functions	253
Decimal Points	254
Add, Subtract, Multiply, and Divide	254
Graphing Trig Functions	255
Final Steps	256

	Complete Code Listing	257
	Chapter Review	260

17. Slot Machine — 261

- Beginning the Project — 262
  - *A Basic GUI* — 262
  - *Spinning the Wheel* — 264
  - *Creating a Delay* — 266
- Win or Lose — 267
- Complete Code Listing — 268
- Chapter Review — 269

18. VB Encryption — 271

- Encryption Basics — 271
- XOR — 272
- Creating the Applications — 273
  - *A Quick GUI* — 273
- The Code — 274
- Complete Code Listing — 276
- Chapter Review — 277

Appendix A — 279

- Standard Naming Conventions — 279

Appendix B — 283

- Internet Resources — 283

Appendix C — 285

- Upgrading to Visual Basic.NET — 285
  - *New Features* — 285
  - *Ugrading Projects* — 285
  - *Upgrade Wizard* — 286
  - *Automatic Code Upgrades* — 286

Variant to Object	286
Integer to Short	287
APIs	287
Newly Introduced Keywords	288
Arrays	289
Default Properties	289
References to Form Controls	289
Forms and Controls	290
Conclusions	290

Appendix D   291
    Virtual-Key Codes   292
    Constant Definitions   294

Appendix E   297
    ASCII Chart   297

Appendix F   301
    Visual Basic Functions and Statements   301
    Visual Basic Functions   301
    Visual Basic Statements   311

Appendix G   317
    API Constants A Through M   317

Appendix H   355
    API Constants N Through Z   355

Appendix I   389
    API Type Declarations   389

Appendix J   399
    Common Properties, Methods and Events   399
    Specifics for Individual Objects and Controls   401

Appendix K	**427**
About the CD-ROM	427
CD Folders	427
System Requirements	427
Installation	428
Index	**435**

Acknowledgments

There are many people who have been involved with the development of this book and because of their hard work and dedication, you are now holding it. First, I'd like to thank everyone at Charles River Media, especially Dave Pallai, for the opportunity to write this book. Your ideas and efforts have been greatly appreciated and it has been a pleasure working with you. I'd also like to individually thank Courtney Jossart, Brenda Lewis and Kelly Robinson for their help with the manuscript.

I'd like to thank my son, Clayton III, who helps me to keep things such as deadlines in perspective and for reminding me that I need to get a little sleep every now and then.

My parents, Clayton & Donna, for leading me down the right path and providing the foundation for everything I have done.

A big thanks to my brother Johnnie for giving me welcome diversions from the computer and helping me to keep what's left of my rapidly deteriorating basketball skills.

The help and advice offered by Mike D. Jones, editor of *Inside Visual Basic,* has been invaluable. Thanks for answering so many of my questions the past couple of years.

Lastly, I'd like to thank my wife Amy for everything she does on a daily basis.

Preface

You can spend a lifetime attending Visual Basic conferences and reading how-to books or magazine articles, but unless you actually write Visual Basic programs, your skill level will never increase.

As the vast majority of people learn better by following examples, this book will focus on developing a complete application in each chapter instead of focusing on a single topic. Not only will this be a more stimulating way for beginners to learn Visual Basic, but intermediate or advanced users can simply skip to a specific chapter in the book that covers a project they are interested in.

While there are many books about programming in general and specifically Visual Basic, most of them tend to focus entire chapters on a particular topic. While learning about ActiveX controls and Text Boxes is a must for any Visual Basic programmer, you shouldn't need to read four or five mind-numbing chapters to learn about them. That's where this book comes in.

With the exception of Visual Basic itself, this book and CD-ROM contain everything you need to complete the projects, as they don't rely on any commercial controls or add-ons. The following list of projects will help you understand what you'll find in this book.

- Developing a Visual Basic Multimedia Player: a program capable of playing a variety of multimedia files including AVI, MIDI, WAV, and CD-Audio.
- Creating an MP3 Player: a complete MP3 player that can play MP3 files and will display the Tag information which contains basic information about the file, including the song name, artist, and year recorded, to name a few.
- Unique Interface Designs: use the Windows API to create a user interface, with irregularly shaped forms and buttons, giving a variety of effects to normally dull Visual Basic forms.

- Pong: you will build a fully functional game of Visual Basic Pong.
- DirectX and DirectDraw: using DirectX technology, you will create a DirectX (.X) model viewer.
- Visual Basic Screen Saver: develop a complete Windows Screen Saver that draws a gradient background and displays a text message at random locations.
- Word Processor: develop a fully functional word processor.
- FTP Client: use built-in controls to create an FTP client capable of downloading files.
- Web Browser: Microsoft's Web Browser control gives us programmatic control over Internet Explorer and using this technology, we create a multiple document interface (MDI) browser.
- Chat Program: a client and server chat program that will function on a single machine, a local network, or across the Internet.
- Animated Desktop Assistant: the Assistant can be used to open applications, visit your favorite web site or read you the current date and time utilizing Text-To-Speech and Voice Recognition Commands.
- Custom ActiveX Control: the control will retrieve the time from Internet servers and return the value to any program it is placed in.
- Paint Program: develop a Visual Basic Paint Program that has a wide range of drawing and painting tools.
- Screen Capture: A complete screen capture utility using a variety of methods.
- Graphing Calculator: A calculator capable of basic math with the ability to graph trigonometry functions.
- Encryption: Basics of encryption techniques using Visual Basic.

1 Introduction to Visual Basic

Although this book is based around developing projects, it is important to have at least a basic understanding of the Visual Basic environment. As a result, this chapter introduces you to Visual Basic and guides you through the creation of your first Visual Basic program. Topics that you'll be exposed to include the basics of the Visual Basic language, variables, the built-in components, the ingredients that make up a basic application and the Integrated Development Environment (IDE).

The vast majority of screen shots and information pertain to the Visual Basic 6 IDE. However, this chapter addresses the differences in the Visual Basic 6 and Visual Basic.NET IDEs and includes information on how to open and convert version 6 projects to .NET.

For those of you already familiar with Visual Basic, you can feel free to move ahead to Chapter 2 where we will begin working on a multimedia and audio CD player. However, if you are a beginner or an intermediate programmer, this chapter will build a solid foundation onto which you can base your future Visual Basic learning.

The Integrated Development Environment

The Visual Basic Integrated Development Environment (IDE) may be the single biggest reason for the vast popularity of Visual Basic. It provides everything you need to develop applications in an easy-to-use-and-learn Graphical User Interface (GUI – pronounced Gooey).

Like many Windows applications, Visual Basic has several ways in which it can be opened. First, and probably the easiest way to access Visual Basic is through the Windows Start menu – the exact path required to access this shortcut is dependent upon your installation and may differ on individual machines. Another option is to create a shortcut on your desktop, which will execute Visual Basic by double-clicking on it. Lastly, because Visual Basic sets up default associations when it is installed, you can also run it by double-clicking on files that end with a vpb (Visual Basic Project) extension.

When you open Visual Basic, you're presented with an opening screen that will appear very much like Figure 1.1. For most of the projects in this book, you will be creating Standard EXE files, so you should click the Open button or just press the Enter key. This standard type of project will create a standard Windows executable program that can be run outside of the IDE.

As you can see in Figure 1.2, the Visual Basic IDE is fundamentally a collection of Menus, Toolbars, and Windows that come together to form the GUI. There are five main Windows that appear in the default Visual Basic IDE, along with the Standard Toolbar, Menu Bar, and Title Bar.

If you have ever used VB5, the version 6 IDE will look very similar to you. In fact, the two versions are so close that you can probably begin working in version 6 without looking over any additional documentation. If you have used Visual Basic version 4 or earlier, the IDE may take some getting used to, as Visual Basic 5 and 6 have been changed to a Multiple Document Interface (MDI) application. If you are new to Visual Basic, the IDE can seem a little daunting at first glance. However, this will quickly fade away as you begin to become comfortable with each part.

Introduction to Visual Basic 3

FIGURE 1.1 The New Project Window is displayed when Visual Basic is started.

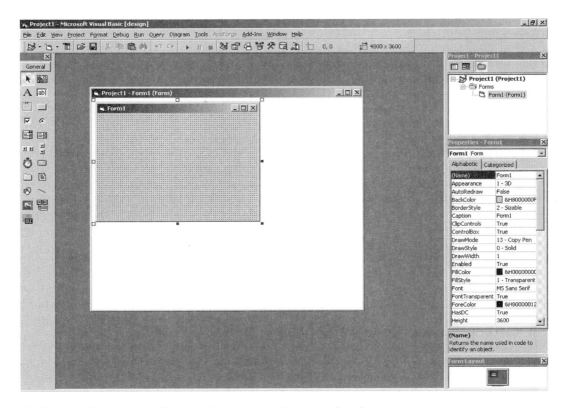

FIGURE 1.2 Visual Basic offers an IDE that is both powerful and easy to use.

The Title Bar and Menu Bar

As you can see in Figure 1.3, the Visual Basic IDE contains a Menu Bar and Title Bar that appears very similar to most Windows applications. The Title Bar serves as a quick reminder of what you are doing inside the IDE. For instance, unless you have changed something, the Title Bar should currently read "Project 1 – Microsoft Visual Basic [design]."

FIGURE 1.3 Menu Bars and Title Bars provide information similar to most Windows programs.

The Menu Bar provides functions that you would expect from any standard Windows application. For instance, the File Menu allows you to Load and Save projects; the Edit Menu provides Cut, Copy, and Paste Commands that are familiar to most Windows users; and the Window Menu allows you to open and close Windows inside the IDE. Each Menu Option works like it would in any other Windows application, so they don't need any real introduction. You shouldn't be overly concerned with all of the options at this time, because we'll be spending some time on them later in this chapter and throughout the book.

THE TOOLBARS

Standard Toolbar

The Standard Toolbar, which is displayed in Figure 1.4, is also comparable to the one found in the vast majority of Windows applications. It provides shortcuts to many of the commonly used functions provided by Visual Basic and along with the Standard Toolbar, Microsoft has provided several additional built-in Toolbars that can make your job a little easier. To add or remove any of the Toolbars, you can right-click on the Menu Bar, or you can select Toolbars from the View Menu.

FIGURE 1.4 Toolbars provide shortcuts to many of the common functions.

Individual Toolbars

The Individual Toolbars include the Debug, Edit and Form Editor Toolbars. The Debug Toolbar, which is visible in Figure 1.5, is utilized for resolving errors in your program and provides shortcuts to commands that are found in the Debug Menu.

FIGURE 1.5 Shortcuts in the Debug Toolbar are helpful for finding errors in your program.

The Edit Toolbar

In Figure 1.6, you will find the Edit Toolbar, which can be useful when you're editing code and setting breakpoints and bookmarks. This Toolbar contains Commands that can be located in the Edit Menu.

FIGURE 1.6 The Edit Toolbar offers a variety of time-saving features.

The Form Editor Toolbar

The Form Editor Toolbar (see Figure 1.7) includes most of the Commands in the Format Menu and is useful only when you're arranging Controls on a Form's surface.

FIGURE 1.7 The Form Editor Toolbar displays Buttons specific to Form editing features.

Whether you decide to display these Toolbars is purely a matter of personal taste, as the functions they provide are generally available in Menu Options. Several factors, such as your screen size and resolution, may make their use impractical.

You can create a custom Toolbar or customize the appearance of the built-in Toolbars by following a couple of steps. First, right-click on any Toolbar, and then select the Customize option. From the Customize Window that appears, click the New button and type a name for the new Toolbar. The name will appear in the Toolbar list, and after making sure that its check box is selected, click the Commands Tab, which displays a list of available Menu Commands. From the list of categories and Commands, select the options you would like to have on your Toolbar. The changes are automatically saved, so continue placing options on the Toolbar until you are finished and then simply close the Window. You can now use your Toolbar like any other.

THE WINDOWS

In addition to the Menus and Toolbars, there are several Windows that you need to become familiar with in order to get a basic grasp of the Visual Basic IDE. The Form Window displays the basic building block of Visual Basic applications. The Code Window is where you enter code for your application. The Toolbox Window displays some of the built-in Visual Basic Controls. You can set properties of the components and Forms with the Properties Window. Lastly, the Project Explorer displays the Objects that make up the project you are working on and you can position and view the Forms with the Form Layout Window.

The Toolbox

The Toolbox, which can be seen in Figure 1.8, is probably the Window that you will become familiar with the quickest, as it provides access to all of the standard Controls that reside within the Visual Basic runtime itself. These Controls, known as intrinsic Controls, cannot be removed from the Toolbox, and include the following options.

FIGURE 1.8 Standard Controls, as well as ActiveX Controls, are displayed in the Toolbox.

- Pointer: The pointer is the only item on the Toolbox that isn't a Control. You can use it to select Controls that have already been placed on a Form.

- PictureBox: You use the Picture Box Control to display images in several different graphics formats such as BMP, GIF, and JPEG among others.

- Label: The Label Control is used to display text information that does not have a need to be edited by an end user. It's often displayed next to additional Controls such as Text Boxes to label their use.

- TextBox: You use Text Box Controls for user input. It may be the most widely used Control.

- Frame: A Frame Control is typically used for containing other Controls and for dividing the GUI. Controls placed within a Frame cannot be displayed outside of it, and if the Frame is moved on the Form, the Controls are moved with it.

- CommandButton: Much like the Text Box Control, Command Button Controls are used for input on almost every Form. They are used as standard buttons for input like OK or Cancel.

- CheckBox: If you need the ability to select True/False or Yes/No, the Check Box Control is the correct Control.

- OptionButton: The Option Button Control is similar to the Check Box Control in that it offers the ability to select an option. However, an Option Button Control is most often used when a group of options exists and only one item can be selected. All additional items are deselected when a choice is made.

- ListBox: The List Box Control contains a list of items, allowing an end user to select one or more items.

- ComboBox: Combo Box Controls are similar to List Box Controls, but they only provide support for a single selection.

- ScrollBars: The HScrollBar and VScrollBar Controls let you create scroll bars but are used infrequently because many Controls provide the ability to display their own Scroll Bars.

- Timer: The Timer Control is an oddity when it is compared to other Controls, in that it isn't displayed at runtime. It's used to provide timed functions for certain events.

- DriveListBox, DirListBox, FileListBox: These Controls can be used individually, but many times are used together to provide dialog boxes (also known as windows in this book) that display the contents of Drives, Directories, and Files.

- Shape, Line: The Shape and Line Controls are simply used to display lines, rectangles, circles, and ovals on Forms.

- Image: You can think of the Image Control as a lighter version of the Picture Box Control, and although it doesn't provide all of the functionality that the Picture Box Control does, it consumes fewer resources. As a result, you should use the Image Control whenever possible.

- Data: The Data Control is a component that allows you to connect one or more Controls on a Form to fields in a database.

- OLE: The OLE (Object Linking and Embedding) Control can host Windows belonging to other executable programs. For instance, you can use it to display a spreadsheet generated by Microsoft Excel or a Word document.

Some of the intrinsic Controls are used more frequently and you are likely to become acquainted with them much faster. The TextBox, Command Button, and Label Controls are used in almost all Visual Basic developed applications. While some Controls are very important, others may provide functionality that can be replaced by far superior Controls. For instance, you probably shouldn't use the Data Control, as it cannot be used with ActiveX Data Objects (ADO) data sources.

Additional Controls, known as ActiveX Controls (sometimes referred to as OCX Controls or OLE custom Controls), provide additional functionality and can be added to the Toolbox for use in a project. These components are provided by many third party companies or may have been provided by Visual Basic itself. Many times, these Controls provide extended functionality that makes them much more powerful than the intrinsic Controls. That said, the built-in varieties offer a few advantages that cannot be overlooked. For instance, if you use a third party Control, you will need to distribute it with your application, whereas the intrinsic Controls are included with the Microsoft Visual Basic runtime file.

Form Window

You need to have a place to assemble your Controls, and this is the function of Forms. As you can see in Figure 1.9, the Forms you work with are displayed inside the Form Designer Window. When they are displayed in this way, you can place and manipulate Controls.

Code Window

Every Form has a Code Window, which is where you write the code for your program. The Code Window can be opened in a variety of ways such as double-clicking on a Form or choosing Code from the View Menu. Figure 1.10 displays a sample Code Window.

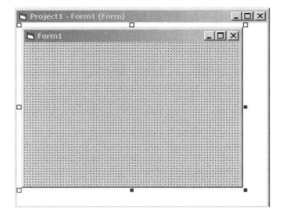

FIGURE 1.9 During development, the Form Designer Window displays the Form you are working on.

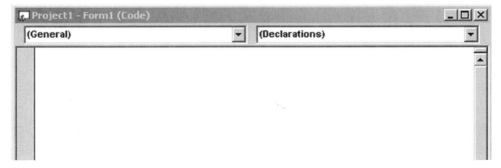

FIGURE 1.10 Visual Basic code is written in the Code Window.

Project Explorer

The Project Explorer can be seen in Figure 1.11 and is provided to help you manage projects. The Project Explorer is a hierarchical tree-branch structure that displays projects at the top of the tree. The components that make up a project, such as Forms, descend from the tree. This makes navigation quick and easy, as you can simply double-click on the part of the project you would like to work on. For instance, if you have a project with several Forms, you can simply double-click the particular Form you want to view.

You can also display the Project Explorer at any time by pressing the F4 key. You can access the Code Window through a shortcut in the Project Explorer. Click on a form to highlight it. Once it is highlighted, you can display the code associated with it by clicking on the Code Window icon that is displayed at the upper left side of the Project Explorer.

The Project Explorer also provides additional functions such as the ability to add new Forms or Code Modules (more on these in later chapters). You can add a Form to a project by right-clicking on an open area of the Project Explorer Window, and selecting Add from the pop-up Context Menu which can be seen in Figure 1.12.

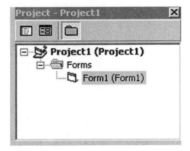

FIGURE 1.11 You'll quickly realize the usefulness of the Project Explorer.

FIGURE 1.12 Pop-up Context Menus make available countless valuable features in the IDE.

Properties Window

The Properties Window is used for the configuration of the Controls you place on a Form, as well as the Form itself. All of the standard Visual Basic Controls have properties, and the majority of ActiveX Controls do as well. As you can see in Figure 1.13, the Window displays the available properties for an individual Control, or the Forms that they are placed on. These properties can be changed as you design an application, or you can alter them in code.

You can display the Properties Window at any time by pressing the F4 key.

Form Layout Window

The Form Layout Window is visible in Figure 1.14. Its only purpose is to allow you to set the position of Forms when they are actually being executed during runtime. The process is very simple. You select the position of the Form by moving it on the small "screen" that represents your desktop. You should keep in mind, however, that the placement of your Form in the Form Designer Window does not affect its position during runtime execution.

FIGURE 1.13 The Properties Window allows you to adjust properties for many Visual Basic Objects.

FIGURE 1.14 You can change the position of executed Forms with the Form Layout Window.

Changes in Visual Basic.NET

Visual Basic.NET is the next generation of Visual Basic and has been completely re-engineered by Microsoft. As you can see in Figure 1.15, Visual Basic.NET introduces some changes to the IDE and new Windows Forms and Web Forms. The product has truly been developed from the ground up and is not just an upgrade to Visual Basic 6.

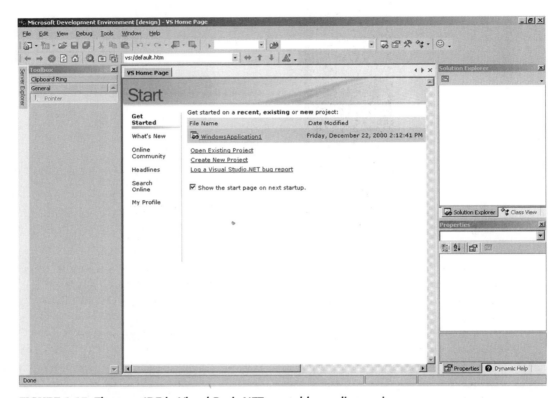

FIGURE 1.15 The new IDE in Visual Basic.NET resembles earlier versions.

Upgrading Version 6 Projects to Visual Basic.NET

Because of the new changes associated with Visual Basic.NET, your code will need to be upgraded before it can be used. Fortunately, the vast majority of time, this is very easy as it happens automatically when you open a Visual Basic 6 project in Visual Basic.NET. An Upgrade Wizard, which can be seen in Figure 1.16, steps you through the upgrade process and creates a new Visual Basic.NET project. The existing Visual Basic 6 project is left unchanged. If you have Visual Basic version 5 projects, it's best to upgrade them to version 6 before moving on to version 7 (VB.NET).

When your project is upgraded, the language is modified for any syntax changes and your Visual Basic 6.0 Forms are converted to Windows Forms.

FIGURE 1.16 The Upgrade Wizard makes it easy to convert version 6 projects to Visual Basic.NET.

Depending on your application, you may need to make minor changes to your code after it is upgraded. Many times this can be necessary because certain features either are not available in Visual Basic.NET, or the features have changed significantly enough to warrant manual changes to the code.

Once your project is upgraded, Visual Basic.Net provides an "upgrade report" to help you make changes and review the status of your project. The items are displayed as tasks in the new Task List Window, so you can easily see what changes are required, and so you can navigate to the code statement simply by double-clicking the task. Many times, the document recommendations simply represent good programming practices, but they also identify the Visual Basic 6 Objects and methods that are no longer supported.

Working with Both Visual Basic 6.0 and Visual Basic.NET

The Visual Basic.NET and Visual Basic 6.0 IDE's can be used on the same computer and can even execute simultaneously. Additionally, applications written and compiled in Visual Basic.NET and Visual Basic 6.0 can be installed and executed on the same computer. Although the projects in this book have been written for Visual Basic 5 or 6, they should work equally well with Visual Basic.NET.

A QUICK PROJECT

Now that you have an understanding of the Visual Basic IDE, you can put the information to work with your first Visual Basic project. The first step to any application is to draw the user interface.

The User Interface

The user interface is probably the first step you'll take when you are developing an application. Start the Visual Basic IDE using one of the methods that were mentioned previously in the chapter and select Standard EXE from the first Window that appears and click the Open button.

The next step is to place Controls on the default Form that appears. There are two separate approaches you can use to do this. First, you can simply double-click on one of the intrinsic Visual Basic Controls that appear in the Toolbox, which will place a single instance of the Control on the Form.

Another way you can place Controls on the Form begins by clicking the Tool in the Toolbox. You then move the mouse pointer to the Form Window, and the pointer changes to a crosshair. Place the crosshair at the upper left corner of where you want the Control to be, press the left mouse button and hold it down while you drag the pointer toward the lower right corner. As you can see in Figure 1.17, when you release the mouse button, the Control is drawn.

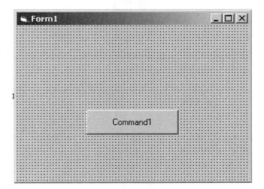

FIGURE 1.17 You can place Controls on a Form in several different ways.

You don't have to place Controls precisely where you want them, as you can move them as Visual Basic provides the necessary tool to reposition them at any time during the development process. To move a Control you have created with either process, click the Object (anywhere on the Object except the edges) in the Form Window and drag it, releasing the mouse button when you have it in the correct location. You can resize a Control very easily as well, by clicking the Object so that it is highlighted and the sizing handles appear. These handles, which can be seen in Figure 1.18, can then be clicked and dragged to resize the Object.

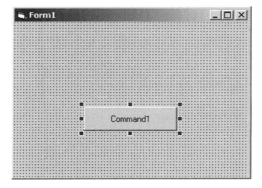

FIGURE 1.18 Handles are useful resizing Objects.

For a first project, you can begin by placing a Text Box and a Command Button on the Form and positioning them so that they look something like Figure 1.19.

FIGURE 1.19 Beginnings of a GUI.

The next step is to double-click the Command Button on the Form, which will bring up the Code Window and leave you something that looks similar to Figure 1.20. It's worth noting that your window will currently only have the Private Sub Command_Click and End Sub lines. Visual Basic automatically creates those lines for all of its intrinsic controls for the most popular used event. In this example, a Command Button's Click Event is its most popular so Visual Basic automatically places the lines in the Code Window.

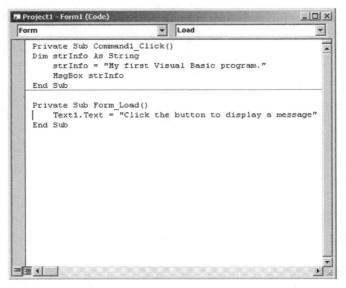

FIGURE 1.20 When you double-click an Object on a Visual Basic Form, it opens an event procedure in the Code Window.

Depending on the way your version of Visual Basic has been set up, Visual Basic may or may not automatically place a line that reads "Option Explicit" when you first open the code window. Option explicit is used by Visual Basic to make sure that you declare all of your variables and is used by default throughout the book. If you don't use the option explicit statement, all undeclared variables are treated as a variant type. You can either type this line at the top of the code window or you can have Visual Basic add it for you by selecting tools | options and clicking the "Require Variable Declaration" option.

Your cursor should be flashing beneath the Private Sub Command1_Click() line. Type the following lines into your application (do not type the Private Sub or

End Sub lines of code for any code shown in this book unless you are instructed to do so):

```
Private Sub Command1_Click()
Dim strInfo As String
   strInfo = "My first Visual Basic program."
   MsgBox strInfo
End Sub
```

From the drop down Menu located at the top left of the Code Window, select Form. A Form _Load event procedure is created for you automatically. Enter the following code and continue reading for an explanation:

```
Private Sub Form_Load()
   Text1.Text = "Click the button to display a message"
End Sub.
```

ON THE CD

The CD-ROM that is included in this book contains all of the sample code for each of the projects we'll create throughout the book. This saves you time and programming mistakes, which will allow you to focus only on the task at hand—learning Visual Basic. It also contains several applications. Please see the CD-ROM for a complet list of applications and compiled applications.

THE CODE EXPLANATION

That's all we need for this application. Although this project is very simple, you are going to be introduced to a few items that you wouldn't necessarily need for this easy of an application, but they are being presented in order to get you started in Visual Basic development. The first of these extras are variables, which are used by Visual Basic to hold information needed by your application. There are only a few simple rules you should keep in mind when you use variables. They should be less than 40 characters; they can include letters, numbers, and underscores (_); they cannot use one of the Visual Basic reserved words (i.e. you cannot name a variable Text); and they have to begin with a letter.

Let's look carefully at what the code you typed does. The "Private Sub Command1_Click()" tells Visual Basic to run this line of code when someone clicks on the Command Button you created called Command1. When they do click the Command Button, the lines of code you typed in are executed. The "End Sub" simply informs Visual Basic that it's time to stop running the code. The Private Sub and End Sub lines were created automatically for you when you double-clicked on the Command1 Button in a previous step.

The Form_Load event was created automatically for you when you selected Form from the drop down Menu in the Code Window. Inside the event, you added a line that sets the Text Box equal to "Click the button to display a message". Like the Command1_Click event, the code is run only when something occurs, which in this case, is the Form being loaded when the program runs.

To use a variable in Visual Basic, you should declare it with the Dim statement. The first line of code that you entered for the Command1_Click event dimensions the variable strInfo as a string. The next line assigns the text string "My first Visual Basic program" to the variable and the final line uses the Visual Basic command MsgBox to display a Message Box containing the text string.

RUNNING THE PROGRAM

You can execute the program from within Visual Basic by clicking the Start button from the Standard Toolbar. You should see a Window that appears something like Figure 1.21.

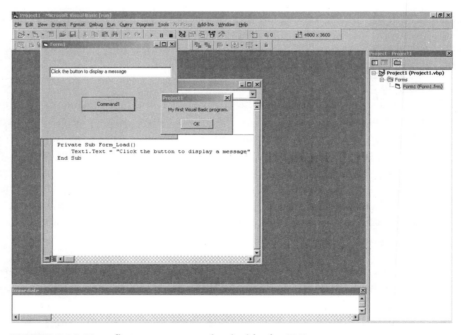

FIGURE 1.21 Your first program running inside the IDE.

You can close it like any Windows program by clicking the X button on the Title Bar or by selecting the Stop button on the Standard Toolbar from within the Visual Basic IDE.

You've created your first program. You can save it if you would like, by choosing Save from the File Menu. When you save a project, it's best to create a new directory in which you can store all the files necessary for the project. In this way, you keep the files in one easy to manage area without the risk of another project corrupting the source code or data.

COMPLETE CODE LISTING

The following code is the complete listing for this chapter:

```
Private Sub Command1_Click()
Dim strInfo As String
  strInfo = "My first Visual Basic program."
  MsgBox strInfo
End Sub

Private Sub Form_Load()
  Text1.Text = "Click the button to display a message"
End Sub
```

CHAPTER REVIEW

During the first chapter, we looked at numerous concepts, many of which might be new. You discovered the Windows, Toolbars, Menus, and Objects that make up the Visual Basic IDE and how to interact with many of them. Visual Basic.NET provides a variety of new features. As a result, Visual Basic 6 projects need to be upgraded by the Upgrade Wizard before they can be used. Lastly, you developed a simple application by using a few intrinsic Controls and some basic code, and then proceeded to run it inside the Visual Basic IDE.

Now that you have some of the basics out of the way, let's move to the next chapter where the real fun begins!

2 Developing a Visual Basic Multimedia Player

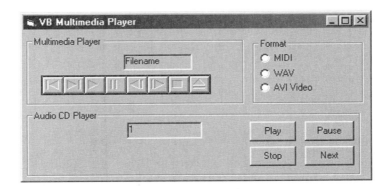

With the introductory chapter out of the way, you can now focus on building the Visual Basic applications that are included in the rest of the chapters, the first of which is a program capable of playing a variety of multimedia files including AVI, MIDI, WAV, and Audio CD. Because this is the first real project in the book, a great deal of time is spent explaining how to do many commonly performed tasks. Later chapters in the book will take considerable time explaining the projects and programming behind them, but will not continue to remind you how to do tasks such as displaying the Components window. If you have trouble, you can always refer back to this chapter as a reference.

PROJECT OVERVIEW

Like the majority of applications, you could use several approaches to complete this program including commercially available or freeware ActiveX controls, DirectX controls, or the Multimedia Control to name a few. While there are a multitude of commercial controls that could be utilized, it is the intention of this book to allow you to develop the applications without the need to buy expensive third party controls or add-ons. As a result, this project will be based around the freely available Microsoft Multimedia Control. The project will introduce you to several aspects of Visual Basic that will help with later chapters.

The User Interface

A great place to begin an application is with the design of the Graphical User Interface (GUI). Start the Visual Basic IDE, using one of the methods that were mentioned previously in Chapter 1. A window, similar to Figure 2.1, will appear from which you can select Standard EXE.

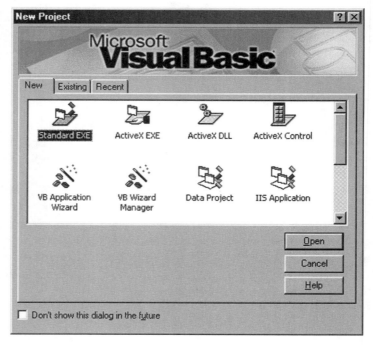

FIGURE 2.1 The Visual Basic New Project Window is displayed on startup.

The next step is to place controls on the default form that appears when you created the project. Because the Multimedia Control is going to be used, you need to add it to the Toolbox. There are several options that you can use to accomplish this, the easiest of which is to use the shortcut key combination CTRL + T. It will display a Components Window that looks like Figure 2.2.

You can also display the components window by right-clicking on the toolbox and selecting components from the pop-up menu.

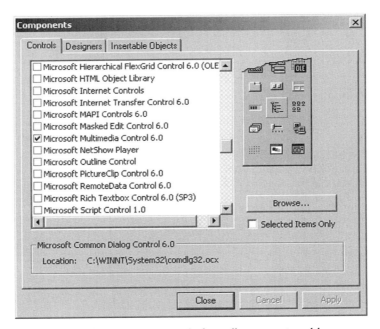

FIGURE 2.2 The Components window allows you to add additional controls.

Scroll down the list of available controls until you find the Multimedia Control. The version number of the Multimedia Control is dependent on the version of Visual Basic you have (i.e. Visual Basic 6 installs Multimedia Control 6.0).

Once you find the appropriate Multimedia Control, select the check box that is located to its left and click the OK button. The Multimedia Control should appear instantly in your Toolbox, which should now look comparable to Figure 2.3.

Next, place and size three Frames on the form in an arrangement similar to Figure 2.4. They will be used as a means to separate the different functions provided in the application.

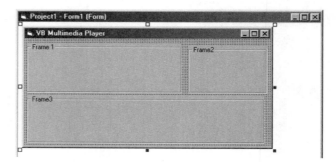

FIGURE 2.4 Frames are used to separate the areas of the application.

FIGURE 2.3 The Multimedia Control is now available for placement in your application.

Now that you have the frames in the correct locations, you need to adjust the Caption properties to reflect their intended use. If the Properties Window, which can be seen in Figure 2.5, is not visible in your Visual Basic IDE, you can display it by pressing the F4 key. Once you locate the Properties Window, click on the top left frame in the Form window and change its Caption property to "Multimedia Player" (the information you enter into the Caption property should not include quotations). The Caption property changes the title that appears on the form when it is executed.

The next step is to continue altering the properties of the other frames so that your Form window looks like Figure 2.6.

FIGURE 2.5 The Properties window provides an easy way to adjust properties.

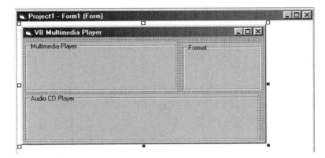

FIGURE 2.6 An interface begins to take shape.

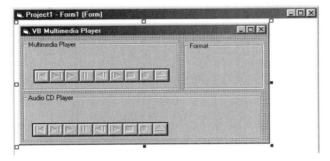

FIGURE 2.7 The Multimedia Controls placed in the frames.

Once the frames are located and the captions are adjusted, you need to place two separate Multimedia Controls on the form — one for the Multimedia Player Frame and one within the Audio CD Player Frame — so that your interface appears like Figure 2.7.

You're using two separate controls in this project as a way to provide additional information for the Multimedia Controls. Normally a single control would suffice for this type of application. One of the controls will use the built-in Multimedia Control interface and will play back AVIs, MIDIs, and WAV files while the other will be used to play Audio CDs and will utilize a series of Command Buttons you add to the frame instead of the built-in variety. The Multimedia Control's built-in interface, which can be seen in Figure 2.8, provides functions that are very similar to a VCR.

FIGURE 2.8 The Multimedia Control contains built-in buttons that work like those on a VCR.

The next step is to rename the Multimedia Controls by using the Properties Window. You can change the Name property for the upper control, which is going to be used to play the various multimedia files, to MMControl, and the Name property for the second control can be changed to MMControlCD.

The name property is at the top of the properties window. Unlike captions, names can be only a single word.

Next, you need to create several Command Buttons that correspond to the functions of an Audio CD Player. The application you are developing will provide basic functions including Play, Stop, Next and Pause. If you choose to do so, once you learn how to program the Multimedia Control, it's easy to add additional options to this application for commands like Previous and Eject. The four command buttons should be placed inside the Audio CD Frame in an arrangement like Figure 2.9. The Command Buttons need to have their names changed to cmdPlay, cmdStop, cmdNextTrack and cmdPause. You can also change their respective Captions to display the appropriate information, i.e. the cmdPlay button should have a caption of Play. Again, you use the Properties Window and when you are finished, your form will now look similar to Figure 2.9.

Inside the Format Frame, you need to place three Option Buttons that will be used to control which type of multimedia file will be played (MIDI, WAV, or AVI Video). You need to be careful to draw the buttons inside the frame by single-clicking the button in the toolbar and then drawing it. You can not double-click a button and then move it from the frame so that it appear to be within the frame as Visual Basic will only think you are positioning it above the frame and not in it. Because you are placing them inside a frame, Visual Basic already understands that only one of the buttons can be selected at any given time and does not require any additional work from you. The option buttons should be renamed to optMIDI, optWAV, and optAVIVideo. Also, their captions should also be changed to MIDI, WAV and AVI Video. The Format Frame will now look like Figure 2.10.

Your GUI is almost finished. You need to add a Label to the Multimedia Player Frame and one to the Audio CD Player Frame. The Labels will be used to display the filename for the Multimedia Player and the Track for the Audio CD Player. Name the labels lblFilename and lblTrack. The lblFilename caption property should also be altered to read Filename and lblTrack should have its caption changed to Track. The resulting GUI should look something like Figure 2.11.

Developing a Visual Basic Multimedia Player 27

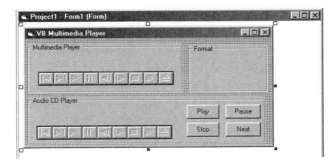

FIGURE 2.9 Command Buttons will provide input for one of the controls.

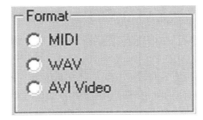

FIGURE 2.10 The Format Frame contains option buttons.

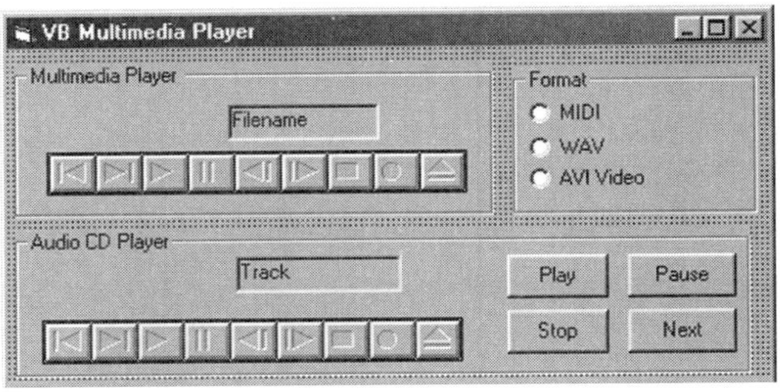

FIGURE 2.11 The final GUI.

PROGRAMMING THE CD PLAYER

Code in a Visual Basic application is divided into small blocks called procedures and are executed when an event occurs. The events can be anything from the user clicking on a button to the application being loaded or unloaded. An event procedure for a particular control combines the control's name, an underscore (_), and the event name. For example, when a form is loaded in Visual Basic, the Form_Load event is triggered. The Form_Load event procedure is a great place to programmatically alter properties of controls or set them to specific values. You can display the Code Window with the cursor inside the Form_Load event procedure by double-clicking on an area of the form that does not have a control. If you do this correctly, you should see something similar to Figure 2.12.

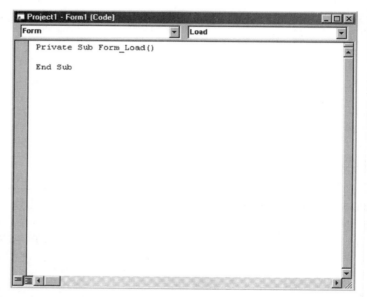

FIGURE 2.12 The Code window with the Form_Load event procedure displayed.

You can also display the code window by clicking on the view code icon in the project explorer window.

NOTE

CD FORM LOAD

Once you are inside the Form_Load event procedure, add the following code. (Remember that you do not need to type the Private Sub or End Sub lines for any code given in this book unless you are instructed to do so.)

```
Private Sub Form_Load()
  MMControl.RecordVisible = False
  MMControlCD.Visible = False
  MMControlCD.DeviceType = "CDAudio"
  MMControlCD.Command = "Open"
End Sub
```

The first line of the above code sets the Record Button to invisible on the MMControl. If you remember, it's this Multimedia Control that will display buttons for playing back the AVI, MIDI, or WAV files. We are not intending this application to record, so it makes sense to set the Visible Property to False.

The second line sets the MMControlCD Visible Property to False, as we intend to control this Multimedia Control with the Command Buttons we placed within the Audio CD Player Frame. You then set the DeviceType of the Multimedia Control to CDAudio and follow it with the Open command. This initializes the MMControlCD control.

Audio CD Commands

The Audio CD Player Commands are going to be sent to the Multimedia Control when the Command Buttons are clicked. The Command_Click event procedures for the buttons will each contain only a single line of code that will send the appropriate command to the Multimedia Control. Because you are already in the Code Window, you can simply click on the Object drop down menu at the top left of the Code window to choose the cmdPlay button. Doing so will automatically change the procedure to the Click procedure, as it is the default for Command Buttons. Inside the event procedure you should enter the following line:

```
Private Sub cmdPlay_Click()
  MMControlCD.Command = "Play"
End Sub
```

This simply sends the Play command to the Multimedia Control.

You can follow this up by creating events for the additional commands Pause, Next Track, and Stop. The following code is all that is required:

```
Private Sub cmdNextTrack_Click()
  MMControlCD.Command = "Next"
End Sub

Private Sub cmdPause_Click()
  MMControlCD.Command = "Pause"
End Sub

Private Sub cmdStop_Click()
  MMControlCD.Command = "Stop"
End Sub
```

You can see how simple it is to program the Multimedia Control. You could easily add functionality to this application by creating additional Command Buttons and command event procedures.

There is one final area that needs to be addressed for the Audio CD Player portion of the application. To be a finished player, the track information needs to be updated as you play the CD. Fortunately, the Multimedia Control provides an easy way to do this as well—the Multimedia Control_Status Update event. This event, which is called every time the status of the control changes, allows you to place code inside the corresponding procedure to set the value of the lblTrack variable equal to the MMControlCD Track Property.

This requires only the following code (select MMControlCD from the Object drop down menu at the top left of the Code window and select StatusUpdate from the Procedure drop down menu at the top right of the Code window):

```
Private Sub MMControlCD_StatusUpdate()
  lblTrack = MMControlCD.Track
End Sub
```

THE MULTIMEDIA PLAYER

The Multimedia portion of the application is a little more involved but is not too difficult. The Format Frame contains several option buttons that will load and play a certain file. The code for this will resemble the Audio CD Player code, but will have some additional commands given. As I previously mentioned, Visual Basic takes care of Option Buttons automatically, and will not let a user select more than one of them at any given time. This considerably eases the programming effort, as you don't have to concern yourself with this.

Much like the Audio CD Player, the Multimedia Player will utilize Click events but instead of Command Buttons being used, it will be the Option Buttons. The basic format for our programming will be to stop any command that has previously been given. We'll then set the Device Type to reflect the appropriate files, set the FileName Property, and finally send the Open command, which will play the file.

Here is the necessary code for each of the Option Button Click event procedures (access each option button through the Object drop down menu at the top left of the Code window; the Click procedure will be automatically selected):

```
Private Sub optAVIVideo_Click()
  MMControlCD.Command = "Close"
  MMControl.DeviceType = "AVIVideo"
  MMControl.FileName = App.Path & "\test.avi"
  MMControl.Command = "Open"
End Sub


Private Sub optMIDI_Click()
  MMControlCD.Command = "Close"
  MMControl.DeviceType = "Sequencer"
  MMControl.FileName = App.Path & "\test.mid"
  MMControl.Command = "Open"
End Sub


Private Sub optWAV_Click()
  MMControlCD.Command = "Close"
  MMControl.DeviceType = "WaveAudio"
  MMControl.FileName = App.Path & "\test.wav"
  MMControl.Command = "Open"
End Sub
```

The first line of each of the Click event procedures closes the MMControlCD so that the Audio CD Player does not play along with the multimedia files. The second line sets the DeviceType property of the MMControl to the appropriate text information. You will notice that the third line sets the FileName Property equal to a hard-coded file. This application could be altered to play files with an Open Dialog Box rather than the hard-coded files. Make sure that you adjust the application to play files that actually exist on your hard drive, as error-handling routines have not been created.

The final Multimedia Control area we need to consider is the StatusUpdate Event of the MMControl. We'll borrow from our earlier work in the MMControlCD StatusUpdate Event and set the value of the lblFilename variable equal to the MMControl FileName Property. Rather than display the entire Path information, we'll use the Right$ command to display only the last eight characters (for each option button, we hard-coded the filename "test" followed by a period and a three letter extension). The following code will produce the desired result (select MMControl from the top left drop down menu and StatusUpdate from the top right drop menu in the Code window):

```
Private Sub MMControl_StatusUpdate()
  lblFilename = Right$(MMControl.FileName, 8)
End Sub
```

ENDING THE APPLICATION

When a Visual Basic program ends, it automatically calls two procedures: Form_QueryUnload and Form_Unload. It first checks the results of QueryUnload and if nothing stops it from exiting, it then continues and calls Unload.

Query Unload

The QueryUnload event offers us the opportunity to ask a user if they really want to exit the application. This isn't the only thing you can do with the QueryUnload event, but it is definitely the most common. In order to do this, we need some way of getting a response from the user. There are several ways we could go about doing this, but the easiest route is to create a Message Box similar to Figure 2.13.

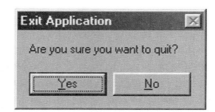

FIGURE 2.13 A message box checks to see if the user really wants to exit.

Once you display the Message Box, you need some way to check the value to determine if the user wants to exit, or if they have decided they want to stay within your application. First, you need to create a variable that will store the information. You can do this with the Dim statement. Type the following code inside the Form_QueryUnload event procedure (select Form from the top left drop down menu and select Query Unload from the top right drop down menu in the Code window):

```
Dim Quit As String
```

Now that you have created the variable Quit, you need to create a Message Box with a Yes/No answer and assign the response to Quit. This can be created by the following code (press the Enter key after you type the underscore):

```
Quit=MsgBox("Are you sure you want to quit?",_
vbYesNo, "Exit Application")
```

The line is very simple. First, we use the equal sign to set Quit equal to the response we receive from the Message Box. If you've never created a Message Box, the format is as follows: MsgBox(prompt[, buttons] [, title] [, helpfile, context]), so in our example, we are prompting the user with the question "Are you sure you want to quit?". We then use the vbYesNo button argument to display Yes and No Buttons on the Message Box and then finish with the "Exit Application" being assigned to the Title Bar of the Message Box which will look very close to Figure 2.13.

Many of the message box options are optional and you would use them only at specific times. In our example, we didn't use helpfile or context.

NOTE

You need to check the results of the Message Box by checking the variable Quit. If it is equal to No, then you should cancel the exit, otherwise you should continue. The following line does this:

```
If Quit=vbNo Then Cancel = True
```

The Form_QueryUnload event procedure is now complete. By passing True back to Cancel if the user selects No from the Message Box, we instruct Visual Basic to cancel the exit. Your final Form_QueryUnload procedure should look like the following:

```
Private Sub Form_QueryUnload(Cancel As Integer, UnloadMode As _
Integer)
```

```
    Dim Quit As String

    Quit = MsgBox("Are you sure you want to quit?", vbYesNo, _
    "Exit Application")

     If Quit = vbNo Then Cancel = True
End Sub
```

You may notice the "_" (underscore) character at the end of the quit = line and the private sub line in the above procedure. This line continuation character is simply used by VB so that you can break up long lines into a series of shorter lines for readability purposes. If you would prefer, you could omit the underscore and have single lines for both entries. Chapters throughout this book utilize the underscore character in the examples.

Unloading the Form

If the user of the application chooses "Yes" from the Message Box, the application will call the Form_Unload event. This event works the opposite of the Form_Load event. You can use the Unload event to stop your application from computing, playing files, loading files, or any other type of tasks. In this example, you can use the Form_Unload event to send the Stop command to both Multimedia Controls. This will keep your application from constantly playing them even after it has exited.

The following code will do this (with Form still selected in the drop down box at the top left of the Code window, select Unload from the drop down box at the top right of the Code window):

```
Private Sub Form_Unload(Cancel As Integer)
   MMControl.Command = "Stop"
   MMControlCD.Command = "Stop"
End Sub
```

TESTING THE APPLICATION

The application should now be ready to execute; however, it is a good idea to save your projects before you execute them. In fact, you should get in the habit of saving them throughout the development process. If you choose File | Save, you will be prompted to save not only the Project, but also any Forms or Modules you have included in your application.

As you can see in Figure 2.14, the Visual Basic IDE offers you an easy way to test your application. Simply by clicking on the Start button (or by pressing F5), you can watch your application execute without having to compile it, which is an obvious time-saver.

When you execute the application, you should see a screen that looks similar to Figure 2.15.

FIGURE 2.14 The Visual Basic IDE.

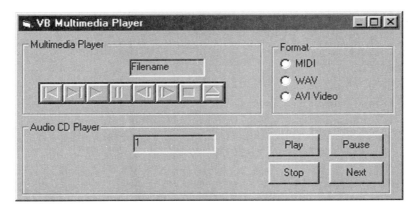

FIGURE 2.15 Your appication should appear something like this.

Because you only had a single form, Visual Basic automatically set it as the default object on startup of your application. If you had more than one form, you would have been required to set the default object by choosing project / project properties from the menu bar.

The application is finished. You can choose to compile an EXE file so that it can be executed outside of the Visual Basic IDE like any Windows application. As I previously mentioned, you can also add commands to the Audio CD Player portion of the program to increase functionality or you could alter the Multimedia Player to allow you to play any multimedia file rather than having it hard coded into the application.

In the following code listing, the procedures may be in a different order than they appear in your code window. Because Visual Basic uses event based programming, it doesn't matter what order they are in so you needn't be alarmed.

Complete Code Listing

The following code is the complete listing for this chapter:

In the Form_QueryUnload event procedure, the line that begins with "Quit = MsgBox..." wraps around to a second line. In reality, this is a single line in Visual Basic, so you shouldn't use a carriage return after "Exit". In this book, carriage returns are obvious because they are indented with the code above and below them whereas a line that is wrapping around will begin at the left margin. Please keep in mind that there are lines like this throughout the book, so if you have any questions, you can refer to the source code on the CD-ROM for further clarification.

```
Option Explicit

Private Sub Form_Load()
  MMControl.RecordVisible = False
  MMControlCD.Visible = False
  MMControlCD.DeviceType = "CDAudio"
  MMControlCD.Command = "Open"
End Sub


Private Sub cmdPlay_Click()
  MMControlCD.Command = "Play"
End Sub


Private Sub cmdNextTrack_Click()
  MMControlCD.Command = "Next"
End Sub
```

```vb
Private Sub cmdPause_Click()
  MMControlCD.Command = "Pause"
End Sub


Private Sub cmdStop_Click()
  MMControlCD.Command = "Stop"
End Sub


Private Sub MMControlCD_StatusUpdate()
  lblTrack = MMControlCD.Track
End Sub


Private Sub MMControl_StatusUpdate()
  lblFilename = Right$(MMControl.FileName, 8)
End Sub


Private Sub optAVIVideo_Click()
  MMControlCD.Command = "Close"
  MMControl.DeviceType = "AVIVideo"
  MMControl.FileName = App.Path & "\test.avi"
  MMControl.Command = "Open"
End Sub


Private Sub optMIDI_Click()
  MMControlCD.Command = "Close"
  MMControl.DeviceType = "Sequencer"
  MMControl.FileName = App.Path & "\test.mid"
  MMControl.Command = "Open"
End Sub


Private Sub optWAV_Click()
  MMControlCD.Command = "Close"
  MMControl.DeviceType = "WaveAudio"
  MMControl.FileName = App.Path & "\test.wav"
  MMControl.Command = "Open"
End Sub
```

```
Private Sub Form_QueryUnload(Cancel As Integer, UnloadMode As Integer)
  Dim Quit As String
  Quit = MsgBox("Are you sure you want to quit?", vbYesNo, "Exit
Application")
  If Quit = vbNo Then Cancel = True
End Sub

Private Sub Form_Unload(Cancel As Integer)
  MMControl.Command = "Stop"
  MMControlCD.Command = "Stop"
End Sub
```

CHAPTER REVIEW

In this chapter, you learned about the Multimedia Control, the properties it provides and how to send commands to it. You also used the Form_QueryUnload Event and a Message Box to prompt the user if they wished to exit the application, as well as Form_Load and Form_Load events. Labels, Frames, Command Buttons and Option Buttons were all used and you set the Properties for each of them—programmatically, and with the Properties Window.

3 Creating an MP3 Player

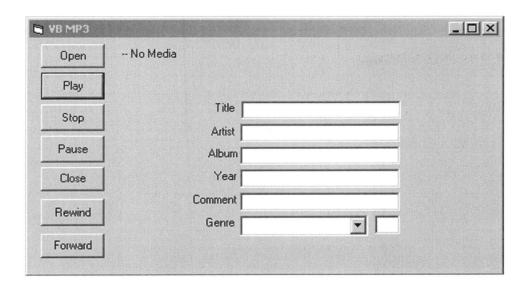

The popularity of MP3 music is unquestioned. There have been several lawsuits directed by the recording industry at companies that aid users in sharing files, which many times are protected under copyright laws. The media attention to the MP3 format is incredible, and with all the publicity, the user base of programs like Napster — one of the many that allow individuals to freely exchange MP3 files — continues to grow nearly exponentially. This is not an endorsement of exchanging copyrighted materials, but rather an acknowledgment that regardless of the outcome of future lawsuits, the format itself is so popular that it will undoubtedly remain in one capacity or another. With this in mind, developing an MP3 player is a very good project for learning Visual Basic programming concepts.

PROJECT OVERVIEW

In this chapter, you'll develop an MP3 player that can play MP3 files and will display the Tag information—information about the file, including the song name, artist, and year to name a few. This project could be based around a number of commercial ActiveX controls, but the goal of this book is to allow you to complete each of the projects without spending any additional money.

Instead of an ActiveX control, you could also remotely control another application from a Visual Basic program. For instance, it would be relatively easy to write a program that would remotely control an existing MP3 player. Again, this option would require you to purchase additional software, which you shouldn't have to do. Another problem with this approach would occur if you decided to distribute your application to other users. Controlling another application would require the user of your MP3 player to purchase another MP3 player. Instead of an ActiveX control or another application, the solution we'll use is the winmm.dll file and Multimedia Control Interface (MCI) commands.

THE PROJECT

To begin our project, we will construct the basic GUI for the MP3 player. Start the Visual Basic IDE and select Standard EXE from the New Project window, which should look similar to Figure 3.1.

FIGURE 3.1 The Visual Basic New Project Window is displayed on startup.

A form is automatically created by Visual Basic. The next step is to alter the Caption property of the form to read "VB MP3" (without quotations). You need to place several controls on the form, one of which is the Microsoft Common Dialog Control. In order to use the Common Dialog Control, you will need to add it to the available components in the Toolbox. There are several options that you can use to accomplish this. The easiest way is to use the shortcut key combination CTRL + T which will display a Components Window that looks similar to Figure 3.2. Scroll through the list of available controls until you find the Common Dialog Control for your particular version of Visual Basic.

If you are unfamiliar with it, the common dialog control provides a standard set of dialog boxes for operations such as opening and saving files, setting print options, and selecting colors and fonts. The control also has the ability to display help by running the windows help engine.

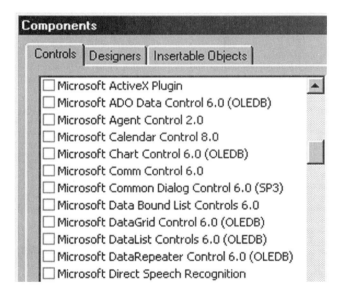

FIGURE 3.2 The Components Window allows you to add additional controls.

FIGURE 3.3 The Common Dialog Control is now available in the Toolbox.

Once you find the appropriate Microsoft Common Dialog Control, select the check box that is located to its left and click the OK button. The control should instantly appear in the toolbox, which will look like Figure 3.3.

Next, place the Common Dialog Control on the form, followed by a series of Command Buttons for opening, stopping, etc. The buttons can be arranged along the left side of the screen going from top to bottom and can be named as follows: cmdOpen, cmdPlay, cmdStop, cmdPause, cmdClose, cmdRewind and cmdForward. Their captions should also be altered to reflect their intended usage; i.e., the cmdOpen button should have its Caption property renamed to Open. When it's finished, the form should look something like Figure 3.4.

As I previously mentioned, in addition to playing MP3 files, we are going to read the Tag information that they contain. In order to display the information, our GUI will need six separate text boxes. You can place and size them in an arrangement similar to Figure 3.5.

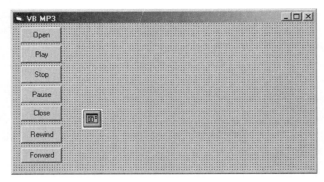

FIGURE 3.4 The interface is beginning to take shape with Command Buttons and the Common Dialog Control in place.

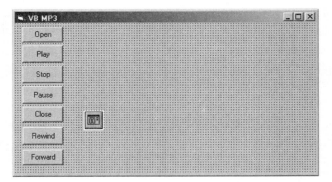

FIGURE 3.5 The Text Boxes will display Tag information for the MP3 player.

The text boxes can be named as follows: txtTitle, txtArtist, txtAlbum, txtYear, txtComment and txtGenreCode. You also need to delete the text for their Text properties. MP3 files store the Genre as a code, so place a combo box using the default name Combo1 with the text for the Text property deleted, which will be used so that you do not have to remember what a particular Genre code relates to. You should also place label controls next to the Text Boxes so that the user of the application can easily differentiate between the contents of the Text Boxes. Their Captions should be altered to reflect the appropriate category: i.e. the label placed beside the txtArtist Text Box should have the Caption property set to Artist. You also need to add a blank label above the text boxes named lblCaption (delete the text in the Caption property; you will set the caption text later in several code statements). The interface should now look similar to Figure 3.6.

 If a specific name isn't given, you can use the standard VB assigned names.

Your final interface should appear something like Figure 3.7.

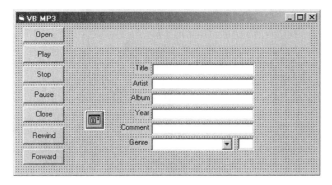

FIGURE 3.6 Text Boxes, the Combo Box and Labels in the appropriate locations.

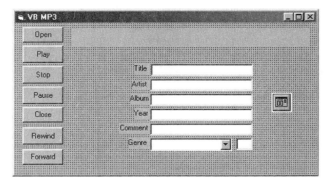

FIGURE 3.7 The final GUI.

SETTING THINGS UP

This application will use several variables to store information. The following list of variables should be entered in the Code Window, which can be displayed by double-clicking the form:

```
Option Explicit
Private Declare Function mciSendString Lib "winmm.dll" Alias
"mciSendStringA" (ByVal lpstrCommand As String, ByVal lpstrReturnString
As String, ByVal uReturnLength As Long, ByVal hwndback As Long) As Long

Dim strFileName As String
Dim blnPlaying As Boolean
Dim GenresTypes As Variant
Dim Temp As Integer

Dim command As String
Dim s As String * 40
```

Although there are several variables, the first line, which begins with Private Declare…, is the most interesting. It's a Windows Application Programming Interface (API) call. For now, you don't need to concern yourself too much with API calls as they will be discussed in detail in the next chapter, but you do need to understand that the mciSendString function is now available to the entire form. The mciSendString function sends a command to a Multimedia Control Interface (MCI) device. The command strings used with this function can perform almost any task necessary for using a multimedia device installed on the computer, and provide a relatively easy way to perform multimedia output operations.

The Form_Load Event is called when the form is first displayed. We'll use this event procedure to set up a few variables and an array that will be used to store the many types of genres. An array is a sequential list of data elements of the same type. Each element has a unique identifying index number. In this example, you can use the Array to store a comma-delimited list of values that are assigned to the Variant GenresTypes variable and later to Combo1. A Variant data type is a special data type that can contain numeric, string, or date data, along with special Null and Empty values.

Enter the following code for the Form_Load Event (select Form from the top left drop down Object menu and Load from the top right drop down Procedure menu):

```
Private Sub Form_Load()
   lblCaption.Caption = " — No Media"
GenresTypes = Array("Blues", "Classic Rock", "Country", _
"Dance",_ "Disco", "Funk", "Grunge", "Hip-Hop", "Jazz", _
"Metal", "New Age", "Oldies", "Other", "Pop", "R&B", _
"Rap", "Reggae", "Rock", "Techno", "Industrial",_
"Alternative", "Ska", "Death Metal", "Pranks", _
"Soundtrack", "Euro -Techno", "Ambient", "Trip -Hop", _
"Vocal", "Jazz Funk", "Fusion", "Trance", "Classical", _
"Instrumental", "Acid", "House", "Game", "Sound Clip", _
"Gospel", "Noise", "AlternRock", "Bass", "Soul", "Punk", _
"Space", "Meditative", "Instrumental Pop", _
"Instrumental Rock", "Ethnic", "Gothic", "Dark Wave", _
"Techno -Industrial", "Electronic", "Pop-Folk", _
"Eurodance", "Dream", "Southern Rock", "Comedy", _
"Cult", "Gangsta", "Top 40", "Christian Rap", "Pop/Funk", _
"Jungle", "Native American", "Cabaret", "New Wave", _
"Psychadelic", "Rave", "Showtunes", "Trailer", "Lo -Fi", _
"Tribal", "Acid Punk", "Acid Jazz", "Polka", "Retro", _
"Musical", "Rock & Roll", "Hard Rock", "Folk", _
"Folk/Rock", "National Folk", "Swing", "Bebop", "Latin", _
"Revival", "Celtic", "Bluegrass", "Avant Garde", "Gothic Rock", _
"Progressive Rock", "Psychedelic Rock", "Symphonic Rock", _
"Slow Rock", "Big Band", "Chorus", "Easy Listening", _
"Acoustic", "Humor", "Speech", "Chanson", "Opera", _
"Chamber Music", "Sonata", "Symphony", "Booty Bass", _
"Primus", "P Groove", "Satire", "Slow Jam", "Club", _
"Tango", "Samba", "Folklore", "Ballad", "Power Ballad", _
"Rhythmic Soul", "Freestyle", "Duet", "Punk Rock", _
"Drum Solo", "A Cappella", "Euro -House", "Dance Hall", _
"Goa", "Drum & Bass", "Club -House", "Hardcore", _
"Terror", "Indie", "BritPop", "Punk", "Polsk Punk", _
"Beat", "Christian Gangsta Rap", "Heavy Metal", _
"Black Metal", "Crossover", "Contemporary Christian", _
"Christian Rock", "Merengue", "Salsa", "Thrash Metal", _
"Anime", "JPop", "Synthpop")


 For Temp = 0 To 146
    Combo1.AddItem GenresTypes(Temp)
 Next Temp
End Sub
```

The first line sets the lblCaption control to inform the user than nothing has been opened. We then assign the various Genres types to the Combo Box named Combo1 by using a For...Next Loop and the AddItem method of the Combo Box. The AddItem method allows you to programmatically add items to a combo box or a list box.

A for...next loop repeats a group of statements a specified number of times. For example, with the following code the variable Y will add 1 to itself in each loop and will end up as a value of 10.

for X = 1 to 10
Y = Y+1
next X

PLAYING THE MP3 FILES

You have created a Command Button called cmdOpen that will be used by the program to open the MP3 file. We simply use the cmdOpen_Click Event which occurs when the button is clicked. First, you need to check to see if a file is already playing, and if so, you should exit without opening a file. To do this, we are using a command called Exit Sub, which allows us to leave an event procedure immediately without continuing down to the more common End Sub. It's often useful if you have determined that the event procedure is not longer needed or in error handling. In our example, we use it to exit the event procedure immediately if we determine a file is playing. However, you should let the user know that they are trying to open the player when a file is already open. You also need to initialize the Common Dialog Control so that it displays only MP3 files. Lastly, you should call the ReadTag subroutine and then use the mciSendString function to open the file.

The last step in the procedure is to add some basic error handling to it. To do this, we use the On Error Goto statement that is built into VB. In our example, an error will occur if the user clicks the Open button and does not actually open a file, i.e., they click cancel. Therefore, we use the error handler to trap the error and then use a Goto statement to inform the program to skip lines until it reaches the line that reads "ErrorHandler:" Many times you would be required to type some specific code after the error handler, but in this case we don't need any code as the only purpose of the error handler is to keep the program from giving us an error.

The following code does all of this: (select cmdOpen from the Object drop down menu)

```
Private Sub cmdOpen_Click()
 On Error GoTo ErrorHandler
 If blnPlaying Then
    MsgBox "Player is Busy!", vbExclamation
    Exit Sub
End If
CommonDialog1.Filter = "MP3 Files|*.mp3"
CommonDialog1.CancelError = True
CommonDialog1.ShowOpen

 If CommonDialog1.FileName = "" Or CommonDialog1.FileName = _
   strFileName Then

Else
  strFileName = CommonDialog1.FileName
  ReadTag

  mciSendString "open " & strFileName & _
  " type MPEGVideo", 0, 0, 0

     lblCaption = strFileName
  End If
ErrorHandler:
End Sub
```

You'll notice the mciSendString syntax, which is very exact and uses quotation marks and includes MPEGVideo as its type. Although it says Video, it is the correct type for MP3 audio files as well.

In the above code, you may have noticed that when we check the value of the boolean variableBe Blnplay, we didn't specify a value. In other words, we didn't type: if Blnplay = "True" then... This is because VB will automatically understand you are trying to determine if a boolean variables is true in the if statements. If you want to check to see if it is False, you would need to type it out as: if BlnPlay = "False" then...

Reading the Tag Information

The next step is to read the Tag information that some MP3 files have. If you are unfamiliar with Tags, they can used to display information about a particular song such as the Artist or the Year in which it was performed. There are several different versions of Tags, with the most popular being ID3. As you can see by Figure 3.8, ID3 Tags stores the various pieces of information in the last 128 bytes of the file.

Audio Data	
Tag	3 Bytes
Title	30 Bytes
Artist	30 Bytes
Album	30 Bytes
Year	4 Bytes
Comment	30 Bytes
Genre	1 Byte
End of File	

FIGURE 3.8 MP3 ID3 Tag Information.

Our first step is to open the file using the command "Open". The position and length of the data we want to read follow the command. Because you know that the Tag information is contained in the last 128 bytes of the file, you can simply calculate the last 128 bytes by taking the FileLength – 127. Reading the data, you can store the string in a variable called "Tag". We will check the Tag variable to determine if it is equal to "TAG"; otherwise, we know that there isn't any Tag information and we can safely exit the procedure. Otherwise, if there is Tag data, we read the data and store it in a custom Type variable. You can continue to read data such as song name, artist, album, year and comment without using any new ideas.

Once you have read the fields, you are forced to handle the Genres category a little differently. An ID3 Tag stores this data as an ASCII character in the MP3 file. To match up the Genre with a category, you must first convert the ASCII character to a number, and then look up that number in the Combo Box.

The following procedure completes our tasks:

```
Private Sub ReadTag()
 On Error Resume Next
 Dim HasTag As Boolean
 Dim Tag As String * 3
 Dim Songname As String * 30
 Dim artist As String * 30
```

```
        Dim album As String * 30
        Dim year As String * 4
        Dim comment As String * 30
        Dim genre As String * 1
        txtTitle = ""
        txtArtist = ""
        txtAlbum = ""
        txtYear = ""
        txtComment = ""
        txtGenreCode = ""
        Combo1.ListIndex = -1
        Open strFileName For Binary As #1
        Get #1, FileLen(strFileName) - 127, Tag
        If Not Tag = "TAG" Then
         Close #1
         HasTag = False
         Exit Sub
        End If
        HasTag = True
        strFileName = strFileName
        Get #1, , Songname
        Get #1, , artist
        Get #1, , album
        Get #1, , year
        Get #1, , comment
        Get #1, , genre
        Close #1
        txtTitle = RTrim(Songname)
        txtArtist = RTrim(artist)
        txtAlbum = RTrim(album)
        txtYear = RTrim(year)
        txtComment = RTrim(comment)
        Temp = RTrim(genre)
        txtGenreCode = Asc(Temp)
        Combo1.ListIndex = txtGenreCode.Text - 1
    Exit Sub
    End Sub
```

The above procedure begins by declaring variables with Dim statements to store the Tag information and then sets the Text Boxes and Combo Box to empty strings. Next, the file is opened for binary reading and checked to see if the "TAG" (in uppercase) string is located in the MP3 file. We use the Get statement to retrieve the information from the file and then close the file. Next, we use Asc(Temp) to convert the chr code stored in Temp to ASCII, which stands for American Standards Code for Information Interchange. ASCII values are used to represent symbols and letters found on US keyboards. Lastly, it removes the trailing spaces of the information by using the RTrim function and sets the Combo Box equal to the appropriate Genre.

THE REST OF THE FUNCTIONS

Once you have the file opened and Tag information displayed, you can use the events associated with clicking any of the command buttons you created for playing, stopping, pausing, etc. The programming that is used in all of the procedures is nearly identical. First, you need to see if a file is already playing by checking the blnPlaying variable, which is of type Boolean. A Boolean variable can display only two values, "True" or False" (could also be thought of as 0 or 1, Off or On). If a file is playing, you can send an MCI code to do the task indicated by the command button (such as close the file). Lastly, you can change lblCaption to reflect the command.

The following lists all of the remaining procedures:

```
Private Sub cmdClose_Click()
If blnPlaying Then
    mciSendString "close " & strFileName, 0, 0, 0
 End If
 blnPlaying = False
 lblCaption.Caption = " —No Media"
 txtTitle = ""
 txtArtist = ""
 txtAlbum = ""
 txtYear = ""
 txtComment = ""
 txtGenreCode = ""
End Sub


Private Sub cmdForward_Click()
If blnPlaying Then

 mciSendString "set " & strFileName & _
 " time format milliseconds", s, 128, O&

 mciSendString "status " & strFileName & _
 " position wait", s, Len(s), 0
```

```vb
   command = "play " & strFileName & _
 " from " & CStr(CLng(s) + 5 * 1000)

 mciSendString command, 0, 0, 0
 blnPlaying = True

 mciSendString "set " & strFileName & _
 " time format frames", 0, 0, 0
 End If
End Sub

Private Sub cmdPause_Click()
  If blnPlaying Then
    mciSendString "pause " & strFileName, 0, 0, 0
    blnPlaying = False
    lblCaption.Caption = strFileName & " —Paused"
  End If
End Sub


Private Sub cmdPlay_Click()
 If strFileName <> "" Then
  mciSendString "play " & strFileName, 0, 0, 0
  blnPlaying = True
  lblCaption.Caption = strFileName & " — Playing"
 End If
End Sub


Private Sub cmdRewind_Click()
 If blnPlaying Then

  mciSendString "set " & strFileName & _
  " time format milliseconds", s, 128, O&

  mciSendString "status " & strFileName & _
  " position wait", s, Len(s), 0

  command = "play " & strFileName & " from _
  " & CStr(CLng(s) - 5 * 1000)

  mciSendString command, 0, 0, 0
  blnPlaying = True
```

```
      mciSendString "set " & strFileName & _
       " time format frames", 0, 0, 0
     End If
   End Sub


   Private Sub cmdStop_Click()
    If blnPlaying Then
      mciSendString "stop " & strFileName, 0, 0, 0
      blnPlaying = False
      lblCaption.Caption = strFileName & " – Stopped"
    End If
   End Sub
```

The only procedure that differs slightly is the cmdPlay_Click procedure, which does not check the status of a playing file. Rather, it only needs to determine if the strFileName variable contains information. If it does, it knows that it must have a file open. If not, it exits the procedure.

Tidying Up

At this point in time, if you were to run the application and exit it while a file is playing, the music would continue to play. You need to add a little code to the Form_Unload event to stop any files that are playing.

This is the code for the event:

```
Private Sub Form_Unload(Cancel As Integer)
  mciSendString "close all", 0, 0, 0
End Sub
```

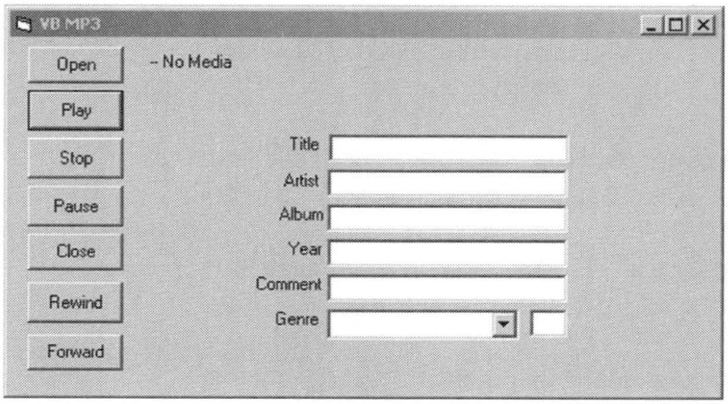

FIGURE 3.9 The final program being executed and displaying MP3 ID3 Tag Information.

Figure 3.9 represents what the finished MP3 player should look like. If you were interested, you could add common functions such as play lists or captions that display time-related information. This project will be used again in the next chapter to add some interesting changes to the vanilla looking GUI.

COMPLETE CODE LISTING

The following code is the complete listing for this chapter:

```
Option Explicit
Private Declare Function mciSendString Lib "winmm.dll" Alias
"mciSendStringA" (ByVal lpstrCommand As String, ByVal lpstrReturnString
As String, ByVal uReturnLength As Long, ByVal hwndback As Long) As Long

Dim strFileName As String
Dim blnPlaying As Boolean
Dim GenresTypes As Variant
Dim Temp As Integer

Private Sub cmdOpen_Click()
 On Error GoTo ErrorHandler
 If blnPlaying Then
  MsgBox "Player is Busy!", vbExclamation
  Exit Sub
 End If
 CommonDialog1.Filter = "MP3 Files|*.mp3"
 CommonDialog1.CancelError = True
 CommonDialog1.ShowOpen
 
 If CommonDialog1.FileName = "" OrCommonDialog1.FileName = strFileName
Then

  Else
   strFileName = CommonDialog1.FileName
   ReadTag
   mciSendString "open " & strFileName &" type MPEGVideo", 0, 0, 0
   lblCaption = strFileName
  End If
ErrorHandler:
End Sub
Private Sub cmdClose_Click()
 If blnPlaying Then
```

```
    mciSendString "close " & strFileName, 0, 0, 0
  End If
  blnPlaying = False
  lblCaption.Caption = " —No Media"
  txtTitle = ""
  txtArtist = ""
  txtAlbum = ""
  txtYear = ""
  txtComment = ""
  txtGenreCode = ""
End Sub

Private Sub cmdForward_Click()
 'Fast Forward
 If blnPlaying Then
  Dim command As String
  Dim s As String * 40
  mciSendString "set " & strFileName & " time format milliseconds", s, 128, 0&
  mciSendString "status " & strFileName & " position wait", s, Len(s), 0
  command = "play " & strFileName & " from " & CStr(CLng(s) + 5 * 1000)
  'The 5 denotes amount of time
  mciSendString command, 0, 0, 0
  blnPlaying = True
  mciSendString "set " & strFileName & " time format frames", 0, 0, 0
  End If
End Sub

Private Sub cmdPause_Click()
  If blnPlaying Then
   mciSendString "pause " & strFileName, 0, 0, 0
   blnPlaying = False
   lblCaption.Caption = strFileName & " —Paused"
  End If
End Sub

Private Sub cmdPlay_Click()
  If strFileName <> "" Then
   mciSendString "play " & strFileName, 0, 0, 0
   blnPlaying = True
   lblCaption.Caption = strFileName & " — Playing"
  End If
End Sub
```

```
Private Sub cmdRewind_Click()
 If blnPlaying Then
  Dim command As String
  Dim s As String * 40
  mciSendString "set " & strFileName & " time format milliseconds", s, 128, 0&
  mciSendString "status " & strFileName & " position wait", s, Len(s),0
  command = "play " & strFileName & " from " & CStr(CLng(s) - 5 * 1000)
  mciSendString command, 0, 0, 0
  blnPlaying = True
  mciSendString "set " & strFileName & " time format frames", 0, 0, 0
 End If
End Sub

Private Sub cmdStop_Click()
 If blnPlaying Then
  mciSendString "stop " & strFileName, 0, 0, 0
  blnPlaying = False
  lblCaption.Caption = strFileName & " — Stopped"
 End If
End Sub

Private Sub Form_Load()
 lblCaption.Caption = " — No Media"

GenresTypes = Array("Blues", "Classic Rock", "Country", _
"Dance",_ "Disco", "Funk", "Grunge", "Hip-Hop", "Jazz", _
"Metal", "New Age", "Oldies", "Other", "Pop", "R&B", _
"Rap", "Reggae", "Rock", "Techno", "Industrial",_
"Alternative", "Ska", "Death Metal", "Pranks", _
"Soundtrack", "Euro -Techno", "Ambient", "Trip -Hop", _
"Vocal", "Jazz Funk", "Fusion", "Trance", "Classical", _
"Instrumental", "Acid", "House", "Game", "Sound Clip", _
"Gospel", "Noise", "AlternRock", "Bass", "Soul", "Punk", _
"Space", "Meditative", "Instrumental Pop", _
"Instrumental Rock", "Ethnic", "Gothic", "Dark Wave", _
"Techno -Industrial", "Electronic", "Pop-Folk", _
"Eurodance", "Dream", "Southern Rock", "Comedy", _
"Cult", "Gangsta", "Top 40", "Christian Rap", "Pop/Funk", _
"Jungle", "Native American", "Cabaret", "New Wave", _
"Psychadelic", "Rave", "Showtunes", "Trailer", "Lo -Fi", _
"Tribal", "Acid Punk", "Acid Jazz", "Polka", "Retro", _
"Musical", "Rock & Roll", "Hard Rock", "Folk", _
"Folk/Rock", "National Folk", "Swing", "Bebop", "Latin", _
```

```
   "Revival", "Celtic", "Bluegrass", "Avant Garde", "Gothic Rock", _
   "Progressive Rock", "Psychedelic Rock", "Symphonic Rock", _
   "Slow Rock", "Big Band", "Chorus", "Easy Listening", _
   "Acoustic", "Humor", "Speech", "Chanson", "Opera", _
   "Chamber Music", "Sonata", "Symphony", "Booty Bass", _
   "Primus", "P Groove", "Satire", "Slow Jam", "Club", _
   "Tango", "Samba", "Folklore", "Ballad", "Power Ballad", _
   "Rhythmic Soul", "Freestyle", "Duet", "Punk Rock", _
   "Drum Solo", "A Cappella", "Euro -House", "Dance Hall", _
   "Goa", "Drum & Bass", "Club -House", "Hardcore", _
   "Terror", "Indie", "BritPop", "Punk", "Polsk Punk", _
   "Beat", "Christian Gangsta Rap", "Heavy Metal", _
   "Black Metal", "Crossover", "Contemporary Christian", _
   "Christian Rock", "Merengue", "Salsa", "Thrash Metal", _
   "Anime", "JPop", "Synthpop")

For Temp = 0 To 146
   Combo1.AddItem GenresTypes(Temp)
  Next Temp
End Sub

Private Sub Form_Unload(Cancel As Integer)
 mciSendString "close all", 0, 0, 0
End Sub

Private Sub ReadTag()
 On Error Resume Next
 Dim HasTag As Boolean
 Dim Tag As String * 3
 Dim Songname As String * 30
 Dim artist As String * 30
 Dim album As String * 30
 Dim year As String * 4
 Dim comment As String * 30
 Dim genre As String * 1

  txtTitle = ""
  txtArtist = ""
  txtAlbum = ""
  txtYear = ""
  txtComment = ""
  txtGenreCode = ""
  Combo1.ListIndex = -1
```

```
    Open strFileName For Binary As #1 ' Read Tag
    Get #1, FileLen(strFileName) - 127, Tag
    If Not Tag = "TAG" Then
     Close #1
     HasTag = False
     Exit Sub
    End If
    HasTag = True
    strFileName = strFileName
    Get #1, , Songname
    Get #1, , artist
    Get #1, , album
    Get #1, , year
    Get #1, , comment
    Get #1, , genre
    Close #1

    txtTitle = RTrim(Songname)
    txtArtist = RTrim(artist)
    txtAlbum = RTrim(album)
    txtYear = RTrim(year)
    txtComment = RTrim(comment)
    Temp = RTrim(genre)
    txtGenreCode = Asc(Temp)
    Combo1.ListIndex = txtGenreCode.Text - 1
    Exit Sub
    End Sub
```

CHAPTER REVIEW

This chapter introduced you to a variety of new concepts including basic API information, the Microsoft Common Dialog Control, the Variant data type, Boolean Data type, and creating and using arrays. You sent commands to the winmm.dll file by using mciSendStrings, which played the MP3 files. You also read Tag information from an MP3 by opening it for binary reading, and then checking the various fields to determine if information was present.

4 Form Effects

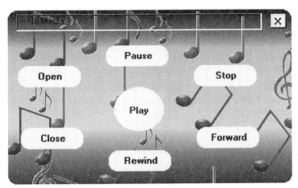 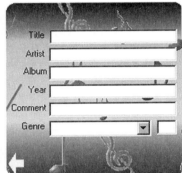

By itself, Visual Basic provides the necessary basic ingredients for Graphical User Interface (GUI) design. Unfortunately, although you can change the color and shapes of items like Forms and Controls, you're severely limited to the variations you can design. In this chapter, you'll create an application that doesn't really have a great deal of functionality but instead focuses on designing a unique interface using the Windows Application Programmer's Interface (API) and Dynamically Linked Libraries (DLLs).

USING THE WINDOWS 32 API

Before you begin working on this project, you need to learn some basic information about the Windows API. I would like to stress the importance of understanding the basics of the API, as you can cause considerable trouble for yourself if you are not careful when you use it.

You can think of the Windows API as a collection of hundreds of ready-made functions and procedures that are available to you inside Visual Basic. The API routines work very similar to Visual Basic's own internal functions and, in fact, you are actually making calls to the API when you use Visual Basic keywords, properties, and methods. However, Visual Basic does a great job of shielding you from the intricacies of the calls.

All the Windows API routines are stored in special files called DLLs, which are simply compiled programs that can be accessed only by other programs and usually have .DLL as their file extension. DLLs cannot be utilized by end users, as they can only be accessed at the programming level. There are several thousand API routines that are available, which can save you time in the programming development cycle.

There are three files that hold most of the API routines, or functions, that you'll call from your Visual Basic applications. The first, USER32.DLL, contains functions related to the control of the program interface such as cursors, menus, and windows. GDI32.DLL is the second, and it provides functions that control output to the screen and other devices. The third most common DLL is KERNEL32.DLL. It gives you access to the internal Windows hardware and software interface, along with memory, file, and directory functions.

If you take a quick glance through the Windows and Windows\System (or WINNT and WINNT\SYSTEM32 for Windows NT or Windows 2000) folders, you'll see these and many more DLL files. The majority of these DLL files are copied to these directories when Windows is installed on your system. You can quickly see that linking to these files is fundamental to the operations of the Windows Operating System.

Because the DLL files are the basis of the Windows Operating System, it's important that you be extremely careful when you use the API, as it is very easy to crash the entire system. Visual Basic does not protect you when you make API calls, so make sure you save your work often when using the API. After all, you wouldn't want your hard work to be lost because of a single misguided call to the API.

If you are afraid that you might forget to save you application at regular intervals, you can check the save before run box in the options | environmental menu. This will provide protection in case of a system crash, which is especially important when you use the windows API.

API and DLL Advantages

You might be asking yourself why you would want to learn about something that can cause such great problems, which is an excellent observation. Visual Basic is very powerful, and you do not have to use the API for the majority of things you do with it. Unfortunately, there are times when Visual Basic does not provide certain functions, or maybe the routines it provides are simply too slow, as is the case with the graphics routines. Still another advantage to using DLLs is that you can use available routines without having to completely rewrite the code. This increases program execution and may make your program easier to develop.

THE FUNDAMENTALS OF API CALLS

In order to access the Windows API, you must first let your program know that you would like to use a particular DLL. Before you call a DLL procedure, it must be declared in your Visual Basic program using the Declare statement. Declare statements go in the General Declarations area of form and code modules (select General from the Object drop down menu in the Code window). The Declare statement informs your program about the name of the procedure, and the number and type of arguments it takes.

The basic call to a DLL function is: Declare Function DLLFunction Lib nameofDLL [(argument list)] As Type. In the example, nameofDLL is a string specifying the name of the DLL file that contains the procedure while Type is the returned value type. A DLL procedure is slightly different: Declare Sub DLLProcecure Lib nameofDLL [(argument list)].

If you place the calls in code modules, you need to preface the Declare statements with the keywords Public or Private, which indicate the procedure scope. This is not a concern in form modules as the only possible use is Private. As a result, you always preface the Declare statement with Private in form modules.

Using the Windows API does not require any tools other than VB itself, but making the calls requires an exact syntax. The code that will be used in the example in this chapter will be completely explained, but if you plan to do any coding with the Windows API on your own, you should pick up a copy of a good book explaining the API in depth.

Because it is so important that the declare statement be correct, you need to familiarize yourself with a program included with visual basic called the api text viewer. This program provides a complete list of declare statements for all api procedures. The viewer can be located by choosing the Visual Basic 6.0 (or 5.0)

tools folder from the start menu, then api text viewer. When you open the application, nothing is displayed. You need to use the file menu to open a file called win32api.txt. It won't tell you what a particular function does, but it will provide the calls in the correct syntax.

PROJECT OVERVIEW

ON THE CD

The project you're going to be developing is based around the MP3 player you worked with in Chapter 3. If you skipped over Chapter 3, you can obtain the source code for the project from the CD-ROM included with this book. The idea is to create a new user interface, with odd-shaped forms and buttons, giving a variety of special effects to our project. This can be accomplished by using regions, which are simply areas on a form that are in the shape of rectangles, ellipses or polygons, and are combined with other regions to form elaborate shapes.

The reason you needed a brief explanation of the Windows API is because VB does not have the ability to directly access regions. On the other hand, the API allows you to create the regions, and returns a handle to it. A handle is simply an "address" that points to something such as a control or a form, or for our specific needs, the regions you are going to create. Before we go on to look at the syntax of the required calls, you can see in Figure 4.1 what our final application will look like.

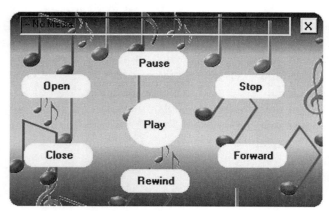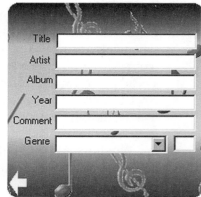

FIGURE 4.1 The final MP3 player interface.

THE API CALLS

Before you look at the code that will make up the application, you can get acquainted with the API calls that you'll be using. As I previously mentioned, you'll be using the API to make regions. The following list details some of the API calls that can be used:

Creating Rectangles

Private Declare Function CreateRectRgn Lib "gdi32" _
(ByVal X1 As Long, ByVal Y1 As Long, ByVal X2 As _
Long, ByVal Y2 As Long) As Long

This creates a rectangularly-shaped region and provides a handle to it. Passing its upper-left and lower-right corners to the function specifies the rectangle defining the region. It's not used in our example, but you could substitute it if you would like.

Creating Ellipses

PrivateDeclare Function CreateEllipticRgn Lib "gdi32" _
(ByVal X1 As Long, ByVal Y1 As Long, ByVal X2 As Long, _
ByVal Y2 As Long) As Long

This creates an elliptically shaped region. The ellipse which forms the region is specified by the bounding rectangle defined by the coordinates passed to the function. It's not used in our example, but you could substitute it if you would like.

Creating Rounded Rectangles

PrivateDeclare Function CreateRoundRectRgn Lib "gdi32" _
(ByVal X1 As Long, ByVal Y1 As Long, ByVal X2 As Long, _
ByVal Y2 As Long, ByVal X3 As Long, ByVal Y3 As Long) As Long

This creates a rectangle like the earlier function, but allows you to specify X3 and Y3 coordinates for creating a rounded rectangle. It's used for all the buttons as well as the form in the example that is created in this chapter.

Create Polygons or Custom Shapes

PrivateDeclare Function CreatePolygonRgn Lib "gdi32" _
(lpPoint As POINTAPI, ByVal nCount As Long, ByVal _
nPolyFillMode As Long) As Long

This allows you to create any shape based on a series of points. It's used to make an arrow effect in the example.

Combining Regions

Once you have created the regions, you can combine them with others and set the window's region to this newly created shape. Here is the API function for it:

Private Declare Function CombineRgn Lib "gdi32" _
(ByVal hDestRgn As Long, ByVal hSrcRgn1 As Long, _
ByVal hSrcRgn2 As Long, ByVal nCombineMode As Long) As Long

This API function combines two or more regions to form a third region. The two regions can be combined using a variety of logical operators, but the region that receives the combined regions must already be a region. You can complete a blank region, to use for this purpose, or you could simply use any of the original regions.

The logical operators for combining regions include:

- RGN_AND: The combined region is the overlapping area of the two source regions.

- RGN_OR: The combined region is all the area contained in either of the two source regions, including any overlap.

- RGN_XOR: The combined region is all of the area contained in either of the two source regions, excluding any overlap.

- RGN_DIFF: The combined region is all the area of the first source region except for the portion also included in the second source region.

- RGN_COPY: The combined region is identical to the first source region. The second source region is ignored.

Initializing the Window

Private Declare Function SetWindowRgn Lib "user32" _
(ByVal hWnd As Long, ByVal hRgn As Long, ByVal bRedraw As Boolean) As Long

This is the function that actually changes the shape of the window. Any part of the form that is outside of the region is not drawn, and thus is invisible to the end user.

SETTING UP THE FORM

If you have not yet done so, open the VB MP3 project you worked on in Chapter 3. Our new GUI will have three new Command Buttons for the extra operations that are needed for closing the program and hiding and showing the Tag information.

You need to create these Command Buttons and place them on the form at exact locations. Because the form will be rounded, the buttons could be placed in an area that is not displayed and would obviously not provide their needed functionality.

The following information will allow you to set the properties for the three new Command Buttons:

cmdCloseForm: Top 110, Left 4734, Width 284, Height 284, Caption X

cmdDisplayTag: Top 2891, Left 4691, Width 575, Height 576, Caption Tag Disp

cmdHideTag: Top 2891, Left 5288, Width 575, Height 575, Caption Tag Hide

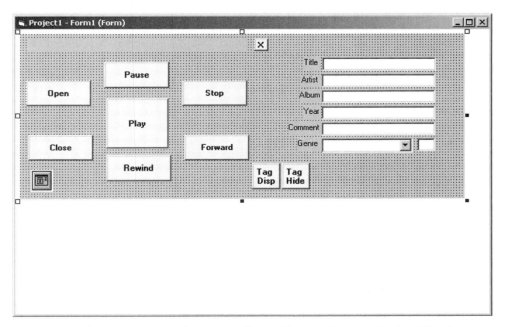

FIGURE 4.2 The new Command Buttons will provide extra features for the MP3 player.

These buttons should look similar to Figure 4.2.

The Command Buttons that were previously on the form can also be arranged to appear similar to the figure and their properties should be adjusted as well. The BackColor property should be changed to yellow for all of the buttons(&H00C0FFFF&) and the following additional properties need to be set:

cmdPause, cmdOpen, cmdStop, cmdForward, cmdRewind, cmdClose: Width 1307, Height 572

cmdPlay: Width 1250, Height 1115

Next, you need to set the form's Border Style to 0-None in the Properties window. The size of the form should also be altered to a Width of 9480 and Height of 3660.

Finally, you need to set the size and position of the labels, combo box and text boxes as follows:

txtTitle: Top 542, Left 6100

txtArtist: Top 903 , Left 6100

txtAlbum: Top 1265, Left 6100

txtYear: Top 1626, Left 6100

txtComment: Top 1988, Left 6100

Combo1: Top 2349, Left 6100

txtGenreCode: Top 2349, Left 8032

lblCaption: Top 90, Left 170, Width 4407, Height 331

The labels that correspond to the text boxes can be placed directly to the left of the text box they are associated with.

Some Declarations

Because the form uses API calls, there are a few extra items that need to be declared. In addition to the API calls, you will also need a couple of new variables as well. These variables will be mentioned once you use them.

You can enter the new following information in the General Declarations section of Form1 (click the View Code button in the Project Explorer, then select General from the Object drop down menu in the Code window):

```
Private Declare Function CreatePolygonRgn Lib "gdi32"_
(lpPoint As POINTAPI, ByVal nCount As Long, ByVal _
nPolyFillMode As Long) As Long

Private Declare Function SetWindowRgn Lib "user32" _
(ByVal hwnd As Long, ByVal hRgn As Long, ByVal bRedraw_
As Boolean) As Long

Private Declare Function CreateRoundRectRgn Lib _
"gdi32" (ByVal X1 As Long, ByVal Y1 As Long, ByVal X2 _
As Long, ByVal Y2 As Long, ByVal X3 As Long, ByVal Y3 _
As Long) As Long_

Private Declare Function CombineRgn Lib "gdi32"_
(ByVal hDestRgn As Long, ByVal hSrcRgn1 As Long,_
ByVal hSrcRgn2 As Long, ByVal nCombineMode As Long)_
```

```
        As Long

        Private Declare Sub ReleaseCapture Lib "user32" ()

        Private Declare Function SendMessage Lib "user32" _
        Alias "SendMessageA" (ByVal hwnd As Long, ByVal wMsg _
        As Long, ByVal wParam As Integer, ByVal lParam As _
        Long) As Long

        Private Type POINTAPI
         X As Long
         Y As Long
        End Type

        Private ButtonRegion As Long
        Private lngRegion As Long
        Private lngRegion2 As Long

        Private Const WM_NCLBUTTONDOWN = &HA1
        Private Const HTCAPTION = 2
```

Changes to the Form_Load Event

The Form_Load event procedure has several additions that need to be made to it. First, you need to set the lblCaption BorderStyle property to 1 so that you can see it more easily on the form. This becomes important once you set the BackStyle property of the lblCaption control to 0 to make it transparent. When you execute the MP3 player, you will not have the ability to move the form by dragging the title bar. Instead, you can use an API call, which will only allow you to move the form if you drag the form itself. As a result, if you do not see lblCaption, you might mistakenly forget it is there and attempt to move the form by dragging it. You need to set all the other label controls to a transparent BackStyle property as well.

The following code can be entered below the previous last line of code for Form_Load but before the End Sub statement in the event:

```
lblCaption.BorderStyle = 1
lblCaption.BackStyle = 0
Label1.BackStyle = 0
Label2.BackStyle = 0
Label3.BackStyle = 0
Label4.BackStyle = 0
Label5.BackStyle = 0
Label6.BackStyle = 0
```

68 Learning Visual Basic Through Applications

In Figure 4.1, you could see the cmdHideTag and cmdDisplayTag buttons that were made to look like arrows. To do this effect, you need to initialize a variable called PTS that will be used to store the points that make up an arrow. The points can be thought of like a connect the dots type of game. As you can see in Figure 4.3, the arrows are very simple and easy to construct.

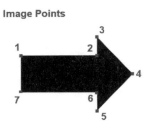

FIGURE 4.3 Creating an array of points allows you to generate an arrow effect.

This will need to be done twice, one for each button and facing in opposite directions. Next, you will use the SetWindowRgn API and the CreatePolygonRgn API to draw the buttons in the arrow shape. With the arrows now drawn, you can set the background of the form to an image of some type represented by the file background.jpg. The following code will accomplish this (add this code below the existing code for Form_Load but above the End Sub statement):

```
Dim PTS(1 To 7) As POINTAPI
  pts(1).X = 5
  pts(1).Y = 15
  pts(2).X = 15
  pts(2).Y = 15
  pts(3).X = 15
  pts(3).Y = 5
  pts(4).X = 25
  pts(4).Y = 20
  pts(5).X = 15
  pts(5).Y = 35
  pts(6).X = 15
  pts(6).Y = 25
  pts(7).X = 5
  pts(7).Y = 25
```

```
cmdDisplayTag.Caption = ""
ButtonRegion = CreatePolygonRgn(pts(1), _
UBound(pts), 1)
SetWindowRgn cmdDisplayTag.hwnd, ButtonRegion, True

pts(1).X = 5
pts(1).Y = 20
pts(2).X = 15
pts(2).Y = 35
pts(3).X = 15
pts(3).Y = 25
pts(4).X = 25
pts(4).Y = 25
pts(5).X = 25
pts(5).Y = 15
pts(6).X = 15
pts(6).Y = 15
pts(7).X = 15
pts(7).Y = 5

cmdHideTag.Caption = ""
ButtonRegion = CreatePolygonRgn(pts(1), _
UBound(pts), 1)
SetWindowRgn cmdHideTag.hwnd, ButtonRegion, True

"\background.jpg")
```

If you recall Figure 4.1, the Command Buttons that control the MP3 functions are made to appear in a rounded rectangular fashion. You can use the CreateRoundRectRgn API along with the SetWindowRgn API to handle the effect (continue adding this code below the code you have just entered):

```
lngRegion = CreateRoundRectRgn(5, 5, 85, 34, 25, 25)
SetWindowRgn cmdOpen.hwnd, lngRegion, True
lngRegion = CreateRoundRectRgn(5, 5, 85, 34, 25, 25)
SetWindowRgn cmdPause.hwnd, lngRegion, True
lngRegion = CreateRoundRectRgn(5, 5, 85, 34, 25, 25)
SetWindowRgn cmdStop.hwnd, lngRegion, True
lngRegion = CreateRoundRectRgn(5, 5, 85, 34, 25, 25)
SetWindowRgn cmdClose.hwnd, lngRegion, True
lngRegion = CreateRoundRectRgn(5, 5, 85, 34, 25, 25)
SetWindowRgn cmdRewind.hwnd, lngRegion, True
lngRegion = CreateRoundRectRgn(5, 5, 85, 34, 25, 25)
SetWindowRgn cmdForward.hwnd, lngRegion, True
```

Lastly, you need to create a circular button for the cmdPlay button and the main part of the form should be adjusted to look like a rounded rectangle. Again, you can use the CreateRoundRectRgn and SetWindowRgn API's to do this (continue with this code below the code you just entered):

```
lngRegion = CreateRoundRectRgn(12, 6, 76, 65, 65, 65)
SetWindowRgn cmdPlay.hwnd, lngRegion, True
lngRegion = CreateRoundRectRgn(1, 1, 360, 230, 25, 25)
SetWindowRgn Me.hwnd, lngRegion, True
```

Moving the Form

With the title bar is removed from the form, you lose the ability to move the form by dragging the title bar with the mouse. Instead, you are forced to use API calls to move the form whenever you click and drag it with the mouse. You can use the MouseDown event to capture any mouse activity and then call a procedure called MoveForm Mouse that contains the API calls.

The following procedures can be added to the project (place this code below the End Sub statement for Private Sub ReadTag (); select Form from the Object drop down menu and MouseDown from the Procedure drop down menu:

```
Private Sub Form_MouseDown(Button As Integer, Shift As Integer, _
X As Single, Y As Single)
 If Button = 1 Then MoveFormMouse
End Sub

Sub MoveFormMouse()
 Dim ReturnVal As Long
 ReleaseCapture
 ReturnVal = SendMessage(Form1.hwnd, WM_NCLBUTTONDOWN,_
 HTCAPTION, 0)
 If ReturnVal = 0 Then
 End If
End Sub
```

FINISHING THE PROJECT

The final procedures that will finish the project use several of the same API calls you have already used — namely, CreateRoundRectRgn and SetWindowRgn — to simulate showing and hiding the Tag information when the cmdDisplayTag and cmdHideTag buttons are clicked. The final step in the project is to exit it with the cmdCloseForm button.

The following procedures can be added to the project immediately below the code you just entered):

```
Private Sub cmdDisplayTag_Click()
 lngRegion = CreateRoundRectRgn(1, 1, 360, 230, 25, 25)
 lngRegion2 = CreateRoundRectRgn(375, 1, 600, 230, 25, 25)
 CombineRgn lngRegion2, lngRegion, lngRegion2, 3
 SetWindowRgn Me.hwnd, lngRegion2, True
 cmdDisplayTag.Visible = False
End Sub

Private Sub cmdHideTag_Click()
 cmdDisplayTag.Visible = True
 lngRegion = CreateRoundRectRgn(1, 1, 360, 230, 25, 25)
 SetWindowRgn Me.hwnd, lngRegion, True
End Sub
Private Sub cmdCloseForm_Click()
 Unload Me
End Sub
```

COMPLETE CODE LISTING

The following code is the complete listing for this chapter as well as Chapter 3:

```
Option Explicit
Private Declare Function mciSendString Lib "winmm.dll" Alias
"mciSendStringA" (ByVal lpstrCommand As String, ByVal lpstrReturnString
As String, ByVal uReturnLength As Long, ByVal hwndback As Long) As Long

Private Declare Function CreatePolygonRgn Lib "gdi32" (lpPoint As
POINTAPI, ByVal nCount As Long, ByVal nPolyFillMode As Long) As Long

Private Declare Function SetWindowRgn Lib "user32" (ByVal hwnd As Long,
ByVal hRgn As Long, ByVal bRedraw As Boolean) As Long

Private Declare Function CreateRoundRectRgn Lib "gdi32" (ByVal X1 As
Long, ByVal Y1 As Long, ByVal X2 As Long, ByVal Y2 As Long, ByVal X3 As
Long, ByVal Y3 As Long) As Long
```

```
Private Declare Function CombineRgn Lib "gdi32" (ByVal hDestRgn As
Long, ByVal hSrcRgn1 As Long, ByVal hSrcRgn2 As Long, ByVal
nCombineMode As Long) As Long

Private Declare Sub ReleaseCapture Lib "user32" ()

Private Declare Function SendMessage Lib "user32" Alias "SendMessageA"
(ByVal hwnd As Long, ByVal wMsg As Long, ByVal wParam As Integer, ByVal
lParam As Long) As Long

Private Type POINTAPI
 X As Long
 Y As Long
End Type

Private ButtonRegion As Long
Private lngRegion As Long
Private lngRegion2 As Long

Private Const WM_NCLBUTTONDOWN = &HA1
Private Const HTCAPTION = 2

Dim strFileName As String
Dim blnPlaying As Boolean
Dim GenresTypes As Variant
Dim Temp As Integer

Private Sub cmdOpen_Click()
 On Error GoTo ErrorHandler
 If blnPlaying Then
  MsgBox "Player is Busy!", vbExclamation
  Exit Sub
 End If
 CommonDialog1.Filter = "MP3 Files|*.mp3"
 CommonDialog1.CancelError = True
 CommonDialog1.ShowOpen

 If CommonDialog1.FileName = "" Or CommonDialog1.FileName =
strFileName Then

 Else
  strFileName = CommonDialog1.FileName
  ReadTag
```

```vb
    mciSendString "open " & strFileName & " type MPEGVideo", _
0, 0, 0
 lblCaption = strFileName
 End If
ErrorHandler:
End Sub

Private Sub cmdClose_Click()
 If blnPlaying Then
  mciSendString "close " & strFileName, 0, 0, 0
 End If
 blnPlaying = False
 lblCaption.Caption = " —No Media"
 txtTitle = ""
 txtArtist = ""
 txtAlbum = ""
 txtYear = ""
 txtComment = ""
 txtGenreCode = ""
End Sub

Private Sub cmdForward_Click()
 'Fast Forward
 If blnPlaying Then
   Dim command As String
   Dim s As String * 40
   mciSendString "set " & strFileName & " time format milliseconds", s, _
128, 0&
   mciSendString "status " & strFileName & " position wait", s, Len(s), 0
   command = "play " & strFileName & " from " & CStr(CLng(s) + 5 * 1000)
'The 5 denotes amount of time
   mciSendString command, 0, 0, 0
   blnPlaying = True
   mciSendString "set " & strFileName & " time format frames", 0, 0, 0
  End If
End Sub

Private Sub cmdPause_Click()
 If blnPlaying Then
   mciSendString "pause " & strFileName, 0, 0, 0
   blnPlaying = False
   lblCaption.Caption = strFileName & " —Paused"
 End If
End Sub
```

```vb
Private Sub cmdPlay_Click()
 If strFileName <> "" Then
  mciSendString "play " & strFileName, 0, 0, 0
  blnPlaying = True
  lblCaption.Caption = strFileName & " — Playing"
 End If
End Sub

Private Sub cmdRewind_Click()
 If blnPlaying Then
  Dim command As String
  Dim s As String * 40
  mciSendString "set " & strFileName & " time format milliseconds", s, 128, 0&
  mciSendString "status " & strFileName & " position wait", s, Len(s), 0
  command = "play " & strFileName & " from " & CStr(CLng(s) - 5 * 1000)
  mciSendString command, 0, 0, 0
  blnPlaying = True
  mciSendString "set " & strFileName & " time format frames", 0, 0, 0
 End If
End Sub

Private Sub cmdStop_Click()
 If blnPlaying Then
  mciSendString "stop " & strFileName, 0, 0, 0
  blnPlaying = False
  lblCaption.Caption = strFileName & " — Stopped"
 End If
End Sub

Private Sub Form_Load()
 lblCaption.Caption = " — No Media"

GenresTypes = Array("Blues", "Classic Rock", "Country", _
"Dance",_ "Disco", "Funk", "Grunge", "Hip-Hop", "Jazz", _
"Metal", "New Age", "Oldies", "Other", "Pop", "R&B", _
"Rap", "Reggae", "Rock", "Techno", "Industrial",_
"Alternative", "Ska", "Death Metal", "Pranks", _
"Soundtrack", "Euro -Techno", "Ambient", "Trip -Hop", _
"Vocal", "Jazz Funk", "Fusion", "Trance", "Classical", _
"Instrumental", "Acid", "House", "Game", "Sound Clip", _
"Gospel", "Noise", "AlternRock", "Bass", "Soul", "Punk", _
"Space", "Meditative", "Instrumental Pop", _
"Instrumental Rock", "Ethnic", "Gothic", "Dark Wave", _
```

```
"Techno -Industrial", "Electronic", "Pop-Folk", _
"Eurodance", "Dream", "Southern Rock", "Comedy", _
"Cult", "Gangsta", "Top 40", "Christian Rap", "Pop/Funk", _
"Jungle", "Native American", "Cabaret", "New Wave", _
"Psychadelic", "Rave", "Showtunes", "Trailer", "Lo -Fi", _
"Tribal", "Acid Punk", "Acid Jazz", "Polka", "Retro", _
"Musical", "Rock & Roll", "Hard Rock", "Folk", _
"Folk/Rock", "National Folk", "Swing", "Bebop", "Latin", _
"Revival", "Celtic", "Bluegrass", "Avant Garde", "Gothic Rock", _
"Progressive Rock", "Psychedelic Rock", "Symphonic Rock", _
"Slow Rock", "Big Band", "Chorus", "Easy Listening", _
"Acoustic", "Humor", "Speech", "Chanson", "Opera", _
"Chamber Music", "Sonata", "Symphony", "Booty Bass", _
"Primus", "P Groove", "Satire", "Slow Jam", "Club", _
"Tango", "Samba", "Folklore", "Ballad", "Power Ballad", _
"Rhythmic Soul", "Freestyle", "Duet", "Punk Rock", _
"Drum Solo", "A Cappella", "Euro -House", "Dance Hall", _
"Goa", "Drum & Bass", "Club -House", "Hardcore", _
"Terror", "Indie", "BritPop", "Punk", "Polsk Punk", _
"Beat", "Christian Gangsta Rap", "Heavy Metal", _
"Black Metal", "Crossover", "Contemporary Christian", _
"Christian Rock", "Merengue", "Salsa", "Thrash Metal", _
"Anime", "JPop", "Synthpop")
   For Temp = 0 To 146
    Combo1.AddItem GenresTypes(Temp)
   Next Temp

lblCaption.BorderStyle = 1
lblCaption.BackStyle = 0
Label1.BackStyle = 0
Label2.BackStyle = 0
Label3.BackStyle = 0
Label4.BackStyle = 0
Label5.BackStyle = 0
Label6.BackStyle = 0

Dim pts(1 To 7) As POINTAPI
 pts(1).X = 5
 pts(1).Y = 15
 pts(2).X = 15
 pts(2).Y = 15
 pts(3).X = 15
 pts(3).Y = 5
 pts(4).X = 25
```

```
pts(4).Y = 20
pts(5).X = 15
pts(5).Y = 35
pts(6).X = 15
pts(6).Y = 25
pts(7).X = 5
pts(7).Y = 25

cmdDisplayTag.Caption = ""
ButtonRegion = CreatePolygonRgn(pts(1), UBound(pts), 1)
SetWindowRgn cmdDisplayTag.hwnd, ButtonRegion, True

pts(1).X = 5
pts(1).Y = 20
pts(2).X = 15
pts(2).Y = 35
pts(3).X = 15
pts(3).Y = 25
pts(4).X = 25
pts(4).Y = 25
pts(5).X = 25
pts(5).Y = 15
pts(6).X = 15
pts(6).Y = 15
pts(7).X = 15
pts(7).Y = 5

cmdHideTag.Caption = ""
ButtonRegion = CreatePolygonRgn(pts(1), UBound(pts), 1)
SetWindowRgn cmdHideTag.hwnd, ButtonRegion, True

Set Me.Picture = LoadPicture(App.Path & "\background.jpg")

lngRegion = CreateRoundRectRgn(5, 5, 85, 34, 25, 25)
SetWindowRgn cmdOpen.hwnd, lngRegion, True
lngRegion = CreateRoundRectRgn(5, 5, 85, 34, 25, 25)
SetWindowRgn cmdPause.hwnd, lngRegion, True
lngRegion = CreateRoundRectRgn(5, 5, 85, 34, 25, 25)
SetWindowRgn cmdStop.hwnd, lngRegion, True
lngRegion = CreateRoundRectRgn(5, 5, 85, 34, 25, 25)
SetWindowRgn cmdClose.hwnd, lngRegion, True
lngRegion = CreateRoundRectRgn(5, 5, 85, 34, 25, 25)
SetWindowRgn cmdRewind.hwnd, lngRegion, True
lngRegion = CreateRoundRectRgn(5, 5, 85, 34, 25, 25)
```

```
  SetWindowRgn cmdForward.hwnd, lngRegion, True

 lngRegion = CreateRoundRectRgn(12, 6, 76, 65, 65, 65)
 SetWindowRgn cmdPlay.hwnd, lngRegion, True

 lngRegion = CreateRoundRectRgn(1, 1, 360, 230, 25, 25)
 SetWindowRgn Me.hwnd, lngRegion, True
End Sub

Private Sub Form_Unload(Cancel As Integer)
 mciSendString "close all", 0, 0, 0
End Sub

Private Sub ReadTag()
 On Error Resume Next
 Dim HasTag As Boolean
 Dim Tag As String * 3
 Dim Songname As String * 30
 Dim artist As String * 30
 Dim album As String * 30
 Dim year As String * 4
 Dim comment As String * 30
 Dim genre As String * 1

 txtTitle = ""
 txtArtist = ""
 txtAlbum = ""
 txtYear = ""
 txtComment = ""
 txtGenreCode = ""
 Combo1.ListIndex = -1

 Open strFileName For Binary As #1
 Get #1, FileLen(strFileName) - 127, Tag
 If Not Tag = "TAG" Then
  Close #1
  HasTag = False
  Exit Sub
 End If
  HasTag = True
 strFileName = strFileName
 Get #1, , Songname
 Get #1, , artist
 Get #1, , album
```

```vb
    Get #1, , year
    Get #1, , comment
    Get #1, , genre
    Close #1
    
    txtTitle = RTrim(Songname)
    txtArtist = RTrim(artist)
    txtAlbum = RTrim(album)
    txtYear = RTrim(year)
    txtComment = RTrim(comment)
    Temp = RTrim(genre)
    txtGenreCode = Asc(Temp)
    Combo1.ListIndex = txtGenreCode.Text - 1
Exit Sub
End Sub

Sub MoveFormMouse()
  Dim ReturnVal As Long
  ReleaseCapture
  ReturnVal = SendMessage(Form1.hwnd, WM_NCLBUTTONDOWN, HTCAPTION, 0)
 If ReturnVal = 0 Then
  End If
End Sub

Private Sub Form_MouseDown(Button As Integer, Shift As Integer, X As Single, Y As Single)
 If Button = 1 Then MoveFormMouse
End Sub

Private Sub cmdDisplayTag_Click()
 lngRegion = CreateRoundRectRgn(1, 1, 360, 230, 25, 25)
 lngRegion2 = CreateRoundRectRgn(375, 1, 600, 230, 25, 25)
 CombineRgn lngRegion2, lngRegion, lngRegion2, 3 'RGN_XOR
 SetWindowRgn Me.hwnd, lngRegion2, True
 cmdDisplayTag.Visible = False
End Sub

Private Sub cmdHideTag_Click()
 cmdDisplayTag.Visible = True
 lngRegion = CreateRoundRectRgn(1, 1, 360, 230, 25, 25)
 SetWindowRgn Me.hwnd, lngRegion, True
End Sub
```

```
Private Sub cmdCloseForm_Click()
 Unload Me
End Sub
```

CHAPTER REVIEW

In this chapter, you leaned how to create window regions and by using several new API functions, create a form with a variety of effects. You used API functions to alter the shape of the form and Command buttons, and you were able to create buttons that were shaped like arrows. Half the form was invisible until you clicked on a button to display the Tag information.

5 Creating a VBPong Game

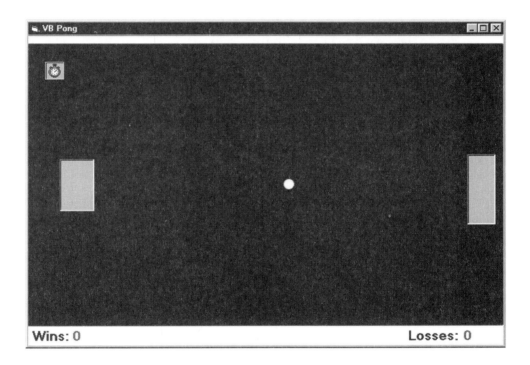

Developing games and basic animation in Visual Basic is one of the tasks that many programmers enjoy. Although Visual Basic is not the ideal environment for creating a 3D shooter like Quake, it excels at simple 2D games. It's not entirely impossible to do 3D games in Visual Basic and with tools such as DirectX (Direct3D is discussed in the next chapter) being supported in Visual Basic, the opportunities for game development in Visual Basic is expanding greatly.

THE PROJECT

In this chapter, you will build a fully functional, albeit simple, game of VBPong, which will be a one player game that could easily be adapted to two players. Along the way, you will learn about a variety of topics including simple 2D graphics and movement.

The GUI

To begin, create a blank application in Visual Basic. As you can see in Figure 5.1, the interface is very simple. On the default form, place two picture boxes so that they are on opposite sides and appear in locations that are close to those shown in Figure 5.1. Change the names of the picture boxes to picPlayer1 and picPlayer2. We'll programmatically alter the additional properties later.

Next, change the background color of the form to any color you would like by using the BackColor property. Keep in mind that darker colors may appear easier to read than lighter colors. Place a shape control in the center of the form and change the properties to the following:

- Name: shpBall

- BackStyle: Transparent

- FillColor: White (or Black if you have chosen a light color for your form background)

- Shape: Circle

- Height: 225

- Width: 225

Once you have the shpBall control in place and its properties adjusted, you need to create two additional shape controls. The first of these needs to be placed at the top of the form and will serve as a top border, while the other needs to be placed at the bottom of the form and will be used as a bottom border. Because they are going to be used as borders, the widths of both shape controls need to extend the full length of the form. You can also change the colors of the border controls to match the color you used earlier for the ball. Figure 5.2 displays all the shapes on the form.

Creating a VBPong Game   83

FIGURE 5.1 The final Pong interface.

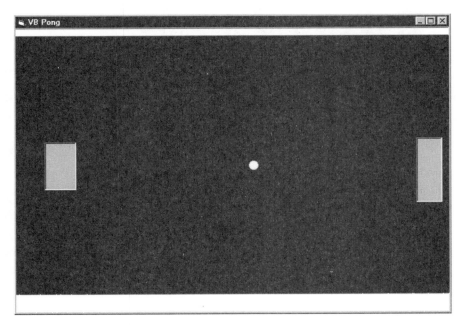

FIGURE 5.2 Shape controls make up most of the GUI.

The GUI is almost finished. You need to place four label controls on the form and alter their names and captions as follows:

Name	Caption
lblWins	Wins:
lblWinsValue	0
lblLosses	Losses:
lblLossesValue	0

You can refer to Figure 5.3 to place the controls in the appropriate locations. The last item you need to position on the form is really not part of the GUI, as it is not displayed at runtime, but this is a good time to place it. You need to position a single timer control on the form and set its Interval property to 1.

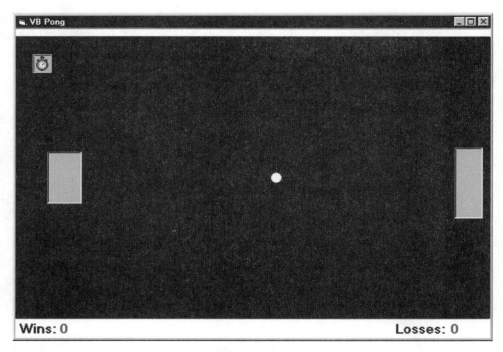

FIGURE 5.3 The labels will be used to display Wins and Losses.

THE CODE

The interface is out of the way, which allows you to begin work on the programming. First, you can add these variables and declarations to the general declarations section:

```
Dim intHorizontalMomentum As Integer
Dim intVerticalMomentum As Integer
Dim intPaddleSpeed As Integer
Dim intOrigPaddleLocation As Integer
Dim wins As Integer
Dim losses As Integer
Dim varDeflectAngle As Variant
Dim intDifficulty As Integer
Const constOrigBallLocX = 5000
Const constOrigBallLocY = 3000
```

Instead of discussing the variables at this point in time, we'll move on to the Form_Load event and will mention the purpose of the variables as we go on.

The Form_Load Event

In the Form_Load Event, you need to record the original position of the player1 paddle so that you can return it to the home position after either side scores a goal. You also need to set the speed at which the ball will move and its initial direction.

The following lines will take care of all of these goals and also introduce you to some of the variables you declared:

```
Private Sub Form_Load ()
  intOrigPaddleLocation = picPlayer1.Top
  intDifficulty = 200
  intHorizontalMomentum = -intDifficulty
  intVerticalMomentum = 0
End Sub
```

There are only a few additional items to take care of in the rest of this event procedure. You need to load graphics to represent the paddles for the two picture box controls (picPlayer1 and picPlayer2) and set a few properties for each. Add the following lines after the previous code above the End Sub line:

```
picPlayer1 = LoadPicture(App.Path & "/paddle.bmp")
picPlayer1.AutoSize = True
picPlayer1.BorderStyle = 0
```

```
picPlayer1.Enabled = False

picPlayer2 = LoadPicture(App.Path & "/paddle.bmp")
picPlayer2.AutoSize = True
picPlayer2.BorderStyle = 0
picPlayer2.Enabled = False
```

The CD-ROM that is included with the book contains all the files including the complete source code to the projects along with files such as the paddle.bmp mentioned above.

After loading the paddle.bmp file, the second line of the above code resizes the picture box so that it equals the exact width and height of the paddle.bmp file. The next line sets the BorderStyle property to none, and lastly, the Enabled property is set to False, which allows the form to maintain focus rather than the Picture Box.

PADDLE MOVEMENT

With the loading of the bitmaps finished, it's time to look for a way to control the movement of the paddles. For our game, the easiest way to do this is to use the Form_KeyDown event, which is why we previously set the picture box controls' Enabled property to False. That allows the form to receive the KeyDown event rather than the picture box controls.

Enter the following code in the Form_KeyDown Sub procedure (with Form selected in the Object drop down menu, select KeyDown in the Procedure drop down menu):

```
Private Sub Form_KeyDown(KeyCode As Integer, Shift As Integer)
  Select Case KeyCode
    Case Is = vbKeyUp
      intPaddleSpeed = -intDifficulty
    Case Is = vbKeyDown
      intPaddleSpeed = intDifficulty
    Case Is = vbKeyF1
      intDifficulty = 100
    Case Is = vbKeyF2
      intDifficulty = 200
    Case Is = vbKeyF3
      intDifficulty = 300
  End Select
End Sub
```

The Select Case statement is used to determine which key is being used and then takes the appropriate action according to the key. If the Up key is pressed, the paddle speed is given a negative value equal to the intDifficulty variable. It is this variable that controls the speed at which the players and ball are moved. If it's the Down key that is pressed, the intPaddleSpeed variable is set to a positive value equal to intDifficulty. The F1, F2 and F3 keys are used to increase or decrease the speed at which the ball and paddles move.

One last thing you need to deal with is stopping the paddle from moving. Once you let up on a key, the paddle will continue moving. To stop this from occurring, you simply use the KeyUp event and set the intPaddleSpeed variable to 0.

The following code will handle this (with Form selected in the Object drop down menu, select KeyUp in the Procedure drop down menu):

```
Private Sub Form_KeyUp(KeyCode As Integer, Shift As Integer)
  intPaddleSpeed = 0
End Sub
```

THE TIMER

The timer that you placed on the form has not been used at this point in time. I saved it for last, as it contains the most code and does most of the work in this example. It's the procedure that handles the collisions, scoring, ball movement, and moving the computer paddle.

BALL MOVEMENT

The first two lines of the procedure should read as follows:

```
Private Sub Timer1_Timer()
  shpBall.Top = shpBall.Top + intVerticalMomentum
  shpBall.Left = shpBall.Left + intHorizontalMomentum
```

These lines are responsible for moving the position of the ball every second by setting the Top and Left properties of the shape control.

When the ball is moved, we need to check to see if it is colliding with anything. Because we cannot check everything at one time, we first check to see if it is colliding with a top or bottom wall. The following code should be entered next:

```
If shpBall.Top + shpBall.Height >= shpWallBottom.Top Then
```

```
      shpBall.Top = shpWallBottom.Top - shpBall.Height
      intVerticalMomentum = -intVerticalMomentum
      Beep
   ElseIf shpBall.Top <= shpWallTop.Top + shpWallTop.Height Then
      shpBall.Top = shpWallTop.Top + shpWallTop.Height
      intVerticalMomentum = -intVerticalMomentum
      Beep
   End If
```

You will notice that the collision is determined by checking the position of the ball relative to the position of the walls. If a collision is made, the momentum is changed to a negative value so that the ball bounces and the Beep command is used to play a simple noise.

The next step is to determine if the ball has collided with a player's paddle. This works much the same way as the wall collision, with the paddle used instead of the wall. Another check needs to be made as, unlike the walls, the paddles can move.

The following code can be entered into the Sub procedure:

```
If shpBall.Left <= picPlayer1.Left + picPlayer1.Width And_
   shpBall.Left >= picPlayer1.Left - picPlayer1.Width Then

   If shpBall.Top + shpBall.Height >= picPlayer1.Top And _
   shpBall.Top <= picPlayer1.Top + picPlayer1.Height Then

      varDeflectAngle = 0.5 * ((picPlayer1.Top + _
      (picPlayer1.Height / 2)) - (shpBall.Top + _
      (shpBall.Height / 2)))

      intVerticalMomentum = intVerticalMomentum + -varDeflectAngle

      Beep

      shpBall.Left = picPlayer1.Left + picPlayer1.Width

   intHorizontalMomentum = -intHorizontalMomentum
 End If
End If

 If shpBall.Left + shpBall.Width >= picPlayer2.Left And _
shpBall.Left <= picPlayer2.Left + _
picPlayer2.Width Then
  If shpBall.Top + shpBall.Height >= picPlayer2.Top And _
shpBall.Top <= picPlayer2.Top +_
 picPlayer2.Height Then
```

```
varDeflectAngle = 0.5 * ((picPlayer2.Top + _
(picPlayer2.Height / 2)) - (shpBall.Top + _
(shpBall.Height / 2)))
  intVerticalMomentum = intVerticalMomentum + -varDeflectAngle
  Beep
  shpBall.Left = picPlayer2.Left - shpBall.Width
  intHorizontalMomentum = -intHorizontalMomentum
End If
End If
```

Earlier in the procedure you created code that would move the ball. You also need to do the same for the computer and player paddles. The following code works along the same principle and moves the paddles by setting their Top property:

```
If intPaddleSpeed <> 0 Then
  picPlayer1.Top = picPlayer1.Top + intPaddleSpeed
End If

If shpBall.Top < picPlayer2.Top Then
  picPlayer2.Top = picPlayer2.Top - 250

ElseIf shpBall.Top > picPlayer2.Top + picPlayer2.Height Then

picPlayer2.Top = picPlayer2.Top + 250

End If
```

Collisions also need to occur if the Player paddles contact the top or bottom walls, or the paddles would move completely out of play. You can check for this collision using the same method you previously used and if a collision occurs, you can simply set the Top position of the paddle below the wall.

The following code handles Player 1 and Player 2 paddle collisions:

```
If picPlayer1.Top <= shpWallTop.Top +shpWallTop.Height Then

   picPlayer1.Top = shpWallTop.Top + shpWallTop.Height

ElseIf picPlayer1.Top + picPlayer1.Height >= shpWallBottom.Top

Then

   picPlayer1.Top = shpWallBottom.Top - picPlayer1.Height

End If
```

```
If picPlayer2.Top <= shpWallTop.Top +shpWallTop.Height Then

picPlayer2.Top = shpWallTop.Top + shpWallTop.Height

ElseIf picPlayer2.Top + picPlayer2.Height >=shpWallBottom.Top

Then

picPlayer2.Top = shpWallBottom.Top - picPlayer2.Height

End If
```

Now that all the collisions have been checked and everything is moving, you need to determine if someone scores and then reset the Player paddles and ball to the beginning locations. You also need to delay the game so a Player realizes they have scored or been scored upon.

SCORING

You can check for scoring by checking the position of the ball and determining if it is less than zero or greater than the width of the form. In both cases, either Player 1 or 2 must have scored a goal. You must stop the movement of the ball, set its position to the original center, and then set the horizontal momentum so that it goes in the direction of the Player who was scored upon.

Lastly, you need to add a point to the Wins or Losses and then pause the game so that you know you scored or were scored upon. You use the Losses or Wins variables to keep track of the score and display it to the screen by changing the lblLossesValue' Caption property or the lblWinsValue' Caption property. You can then use a Message Box to let the user know that they won or lost the game.

The following code finishes the Sub procedure:

```
If shpBall.Left <= 0 Then
  shpBall.Left = constOrigBallLocX
  shpBall.Top = constOrigBallLocY
  picPlayer1.Top = intOrigPaddleLocation
  picPlayer2.Top = intOrigPaddleLocation
  intHorizontalMomentum = -intDifficulty
  intVerticalMomentum = 0
  losses = losses + 1
  lblLossesValue = losses
  MsgBox "The Computer has defeated you!",_
```

```
            vbOKOnly, "Game Over"
            Me.Refresh

        ElseIf shpBall.Left > Form1.Width Then
          shpBall.Left = constOrigBallLocX
          shpBall.Top = constOrigBallLocY
          picPlayer1.Top = intOrigPaddleLocation
          picPlayer2.Top = intOrigPaddleLocation
          intHorizontalMomentum = intDifficulty
          intVerticalMomentum = 0
          wins = wins + 1
          lblWinsValue = wins
          MsgBox "You have defeated the Computer!",_
            vbOKOnly, "Game Over"
          Me.Refresh
        End If
```

COMPLETE CODE LISTING

The following code is the complete listing for this chapter:

```
Option Explicit

Dim intHorizontalMomentum As Integer 'horizontal momentum
Dim intVerticalMomentum As Integer 'vertical momentum
Dim intPaddleSpeed As Integer 'the speed of the players paddle
Dim intOrigPaddleLocation As Integer
Dim wins, losses As Integer
Dim varDeflectAngle As Variant
Dim intDifficulty As Integer
Const constOrigBallLocX = 5000 'Original Value
Const constOrigBallLocY = 3000 'Original Value

Private Sub Form_Load()
  intOrigPaddleLocation = picPlayer1.Top 'starting spot for Player1
  intDifficulty = 200 'Set to Medium Speed
  intHorizontalMomentum = -intDifficulty 'Set Movement To Left
  intVerticalMomentum = 0 'Set Movement to 0

  picPlayer1 = LoadPicture(App.Path & "/paddle.bmp")
  picPlayer1.AutoSize = True
  picPlayer1.BorderStyle = 0 'None
```

```vb
    picPlayer1.Enabled = False 'Do Not Allow to Have Focus
    
    picPlayer2 = LoadPicture(App.Path & "/paddle.bmp")
    picPlayer2.AutoSize = True
    picPlayer2.BorderStyle = 0
    picPlayer2.Enabled = False
End Sub

Private Sub Form_KeyDown(KeyCode As Integer, Shift As Integer)
 Select Case KeyCode
  Case Is = vbKeyUp
 intPaddleSpeed = -intDifficulty
  Case Is = vbKeyDown
 intPaddleSpeed = intDifficulty
  Case Is = vbKeyF1
 intDifficulty = 100
  Case Is = vbKeyF2
 intDifficulty = 200
  Case Is = vbKeyF3
 intDifficulty = 300
 End Select
End Sub

Private Sub Form_KeyUp(KeyCode As Integer, Shift As Integer)
 intPaddleSpeed = 0 'Set paddle to not moving
End Sub

Private Sub Timer1_Timer()
 'Moving the Ball
 shpBall.Top = shpBall.Top + intVerticalMomentum
 shpBall.Left = shpBall.Left + intHorizontalMomentum
 
 'Collision with top or bottom wall
  If shpBall.Top + shpBall.Height >= shpWallBottom.Top Then
     shpBall.Top = shpWallBottom.Top - shpBall.Height
     intVerticalMomentum = -intVerticalMomentum
     Beep
  ElseIf shpBall.Top <= shpWallTop.Top + shpWallTop.Height Then
     shpBall.Top = shpWallTop.Top + shpWallTop.Height
     intVerticalMomentum = -intVerticalMomentum
     Beep
  End If
```

```
'Ball Collision with Player 1 Paddle
If shpBall.Left <= picPlayer1.Left + picPlayer1.Width And shpBall.Left
>= picPlayer1.Left - picPlayer1.Width Then
If shpBall.Top + shpBall.Height >= picPlayer1.Top And shpBall.Top <=
picPlayer1.Top + picPlayer1.Height Then
varDeflectAngle = 0.5 * ((picPlayer1.Top + (picPlayer1.Height / 2)) -
(shpBall.Top + (shpBall.Height / 2)))
'calculate the deflection angle
intVerticalMomentum = intVerticalMomentum + -varDeflectAngle
Beep
shpBall.Left = picPlayer1.Left + picPlayer1.Width
intHorizontalMomentum = -intHorizontalMomentum 'deflect the ball
End If
End If

'Ball Collision with Player 2 Paddle
If shpBall.Left + shpBall.Width >= picPlayer2.Left And shpBall.Left <=
picPlayer2.Left + picPlayer2.Width Then
If shpBall.Top + shpBall.Height >= picPlayer2.Top And shpBall.Top <=
picPlayer2.Top + picPlayer2.Height Then
varDeflectAngle = 0.5 * ((picPlayer2.Top + (picPlayer2.Height / 2)) -
(shpBall.Top + (shpBall.Height / 2)))
intVerticalMomentum = intVerticalMomentum + -varDeflectAngle
Beep
shpBall.Left = picPlayer2.Left - shpBall.Width
intHorizontalMomentum = -intHorizontalMomentum
End If
End If

'Paddle is Moving
If intPaddleSpeed <> 0 Then
picPlayer1.Top = picPlayer1.Top + intPaddleSpeed
End If

'Moving Computer Paddle
If shpBall.Top < picPlayer2.Top Then
 picPlayer2.Top = picPlayer2.Top - 250
ElseIf shpBall.Top > picPlayer2.Top + picPlayer2.Height Then
 picPlayer2.Top = picPlayer2.Top + 250
End If

'Paddle Collide with Wall
If picPlayer1.Top <= shpWallTop.Top + shpWallTop.Height Then
 picPlayer1.Top = shpWallTop.Top + shpWallTop.Height
```

```
    ElseIf picPlayer1.Top + picPlayer1.Height >= shpWallBottom.Top Then
    picPlayer1.Top = shpWallBottom.Top - picPlayer1.Height
    End If

    If picPlayer2.Top <= shpWallTop.Top + shpWallTop.Height Then
     picPlayer2.Top = shpWallTop.Top + shpWallTop.Height
    ElseIf picPlayer2.Top + picPlayer2.Height >= shpWallBottom.Top Then
     picPlayer2.Top = shpWallBottom.Top - picPlayer2.Height
    End If

'Check if someone has scored
If shpBall.Left <= 0 Then
 'If scored, set ball and paddles to their origional location
 shpBall.Left = constOrigBallLocX
 shpBall.Top = constOrigBallLocY
 picPlayer1.Top = intOrigPaddleLocation
 picPlayer2.Top = intOrigPaddleLocation
 intHorizontalMomentum = -intDifficulty
 intVerticalMomentum = 0
 losses = losses + 1
 lblLossesValue = losses
 MsgBox "The Computer has defeated you!", vbOKOnly, "Game Over"
 Me.Refresh

 ElseIf shpBall.Left > Form1.Width Then
 shpBall.Left = constOrigBallLocX
 shpBall.Top = constOrigBallLocY
 picPlayer1.Top = intOrigPaddleLocation
 picPlayer2.Top = intOrigPaddleLocation
 intHorizontalMomentum = intDifficulty
 intVerticalMomentum = 0
 wins = wins + 1
 lblWinsValue = wins
 MsgBox "You have defeated the Computer!", vbOKOnly, "Game Over"
 Me.Refresh
 End If
End Sub
```

CHAPTER REVIEW

In this chapter, you built a game of VBPong and in doing so, learned about basic movement and a very simple way to do collision detection. You could easily adapt this project in a variety of ways, such as adding an option that would allow a second person to play. You were introduced to the timer control and used it to set up a game loop for moving the objects around the screen.

6 A 3D Model Viewer

Prior to DirectX version 7, Visual Basic programmers did not have direct access to the powerful libraries developed by Microsoft. DirectX 8 continues to support Visual Basic, and with components such as DirectDraw, DirectSound, DirectInput and Direct3D, programmers interested in multimedia or game programming will find the tools a necessity.

PROJECT OVERVIEW

ON THE CD

In this chapter, you will be introduced to the Retained Mode of Direct3D and will build an application that will display a DirectX binary model (.x extension) and rotate around the x-axis. The CD-ROM contains the complete Microsoft DirectX8 Software Developers Kit (SDK). In order to run any samples in this chapter, you must first install the SDK.

WHAT IS DIRECT3D?

Direct3D is a single aspect of DirectX, which in turn is a collection of high performance APIs written by Microsoft to produce sound, graphics animation, and multimedia. Direct3D is the three-dimensional graphics part of DirectX.

In versions prior to DirectX 7.0, DirectX was only available to VB programmers via third party controls or Type libraries. As a result, most VB programmers simply ignored it. Beginning with version 7.0, Microsoft started to include a Type library written specifically for VB programmers.

Advantages of Direct3D

While there are a plethora of advantages to Direct3D, it's important to understand the most basic of these. We'll look at a couple of them. First, Direct3D provides transparent access to different hardware devices without the need to understand the intricacies of the different chipsets. As a result, applications become independent of the hardware and allow the developer to immediately know that all of the drivers supporting Direct3D will support a certain set of instructions and capabilities. Applications developed using such features will work on all hardware platforms.

Direct3D also provides a standard programming interface for 3D applications, which results in applications that can be developed much easier and faster. Lastly, Direct3D provides some basic protection if a hardware platform does not support a certain feature, by substituting a software equivalent implementation of the intended feature. Thus, the application can simply detect the hardware capabilities and use them; otherwise, it can render in Software Mode. Obviously, software substitutions do make applications run more slowly, but that's better than not running at all.

The SDK

Before you begin the tutorial, it's important to realize the vast amount of information available to you in the DirectX SDK. The following lists some specific areas that you should visit to become more familiar with DirectX and Direct3D. The following list assumes you have installed the software to drive C using the default mssdk installation path. If you installed the SDK to another directory or hard drive, you will need to change the drive letter and root directory.

C:\mssdk\DXReadme.txt: A file that is particularly useful for navigating the rest of the CD.

C:\mssdk\doc\DirectX8\directX_vb.chm: This is a compiled Help file containing information about DirectX 8 – the most important entry on the entire CD-ROM.

C:\mssdk\samples\Multimedia\VBSamples: This directory contains Visual Basic samples utilizing all the DirectX components.

Direct3D Coordinate System

There are several ways you might arrange the X, Y, and Z coordinate axes in a given three-dimensional space. Direct3D uses a system like the one that appears in Figure 6.1, where the X- and Y-axes are parallel to the computer monitor and the Z-axis is going into the screen.

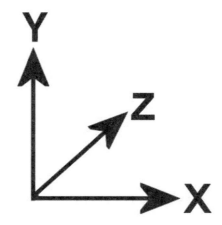

FIGURE 6.1 The coordinate system used by Direct3D.

With Figure 6.1 as a reference, if you tried to imagine your position when you are sitting at your computer desk, relative to the point X0, Y0, Z0 in a Direct3D space, you would be at some point in space with a positive Z coordinate.

DIRECT3D MODES

DirectX 8 provides a single mode, called Immediate Mode, to programmers. This mode is a high-level interface to Direct3D that can be cumbersome and confusing to a beginner. Previous versions of DirectX offered the option of using Retained Mode, which was a low-level implementation. It was much easier to access and understand, especially for someone new to 3D programming. Taking this into consideration, the sample being developed in this chapter will utilize Retained Mode, which can still be accessed by taking advantage of the backward compatibility of DirectX. As a result, you'll be using Direct3D version 7.

If you don't have any experience with Direct3D, Retained Mode is definitely the easiest place to start. A great deal of the more difficult stuff is hidden from you and what's left is a much easier to use Direct3D implementation. What you learn in Retained Mode can also act as a stepping stone for future work.

There is a tradeoff that occurs with the use of Retained Mode. Application frame rates will be much lower with Retained Mode. This would cause a problem if you were developing a large project, but if you are after quick and easy development, Retained Mode is the way to go.

The final disadvantage of Retained Mode is the fact that Microsoft has discontinued it, which was already mentioned. It remains to be seen if Microsoft will continue supporting it in future editions, although there are some persistent rumors that it will be making its way back into DirectX 9 or 10. It's an interesting thought, although I wouldn't bet on it. My suggestion would be to learn Retained Mode, and then move on to Immediate Mode.

3D VOCABULARY

For our sample, you must understand some basic 3D vocabulary. The following are very simple explanations of some of the basics.

- Camera: To do something in 3D space, you have to have a way to represent where you are. This is the sole responsibility of the Camera.

- Rotation: Rotating an object in a 3D world works very similar to a tire rotating on a car axis. Simply stated, rotation is the movement of an object around a coordinate that is referenced to the whole screen but NOT on the local object.

- Scenes: Scenes are comprised of the many objects that you would like to display, including lights, 3D objects, and backgrounds.

- Render Modes: To display the 3D information, you must render it in 2D. There are three main Render Modes: Wireframe, Solid, and Gourad. Wireframe displays only a set of lines and is the simplest. Solid Rendering fills the spaces between the lines with color while Gouraud adds shading to enhance the appearance of objects.

- Lights: A light simply does what you would expect it to. If you do not place a light in a scene, you will see nothing.

PROJECT FRAMEWORK

- The following sections describe the different aspects in the program:

- Declarations: Declares all variables and DirectX.

- Form_Load: Calls all other routines to Initialize DirectX.

- InitializeDirectX: Initializes DirectDraw and Direct3D.

- InitializeScene: Initializes Cameras, objects, etc.

- RenderLoop: Repeatedly draws objects.

- Form_Click: Ends the applications

DECLARATIONS

There are several declarations that you need for this application. To begin, create a new project in Visual Basic and enter the following declarations in the general declarations section:

```
Dim DX As New DirectX7
Dim DD As DirectDraw4
Dim RM As Direct3DRM3
```

The first line is particularly important as it is from the DirectX object that you set up DirectDraw and Direct3D. Notice that New is used after DX which simply tells Visual Basic that you want to put aside memory to use for the DirectX object you are creating.

The next step is to add DirectDraw declarations, which sets the surfaces that you draw to and see. The following four entries can be added:

```
Dim PrimarySurface As DirectDrawSurface4
Dim BackSurface As DirectDrawSurface4
Dim DDPrimarySurfDescript As DDSURFACEDESC2
Dim DDCapsBack As DDSCAPS2
```

After the DirectDraw declarations, you need to add the Direct3D Retained Mode declarations. The following declarations set the devices, viewport, frames, lights, and shadows. Basically they set up the entire Direct3D scene. Enter these after the DirectDraw entries:

```
Dim rmDevice As Direct3DRMDevice3
Dim rmViewport As Direct3DRMViewport2
Dim rootFrame As Direct3DRMFrame3
Dim lightFrame As Direct3DRMFrame3
Dim cameraFrame As Direct3DRMFrame3
Dim objectFrame As Direct3DRMFrame3
Dim light As Direct3DRMLight
Dim shadow_light As Direct3DRMLight
Dim shadow As Direct3DRMShadow2
Dim meshBuilder As Direct3DRMMeshBuilder3
```

Four of the final five variables are used to track the number of frames that are being displayed, while the last one, blnIsRunning, is used to stop the main rendering loop once program execution is halted. Add these final variables to the General Declarations section after your last entry:

```
Dim lngTimeLast As Long
Dim intFramesCompleted As Integer
Dim strFramesText As String
Dim lngTick As Long
Dim blnIsRunning As Boolean
```

Form_Load Event

When the form loads, it calls the other routines, which in turn do the work. The following code can be entered in the Code Window:

A 3D Model Viewer

```
Private Sub Form_Load()
 InitializeDirectX
 InitializeScene
 Me.Show
 RenderLoop
End Sub
```

In the above code, the "me" keyword behaves like an implicitly declared variable. In other words, it is much easier to type me.show instead of form1.show.

InitializeDirectX and InitializeScene are event procedures we are going to create to prepare DirectDraw and Direct3D Retained Mode for execution. We'll take a closer look at the actual procedures in the next section. After we call those procedures, the program then displays the form and makes its final call to RenderLoop, which is the main program loop for displaying the .X file.

Initialize DirectDraw and Direct3D

The InitializeDirectX procedure can be fairly long and complicated if this is your first exposure to 3D. You need to begin the procedure by initializing DirectDraw and getting it ready for full screen display. You can also set up the DirectDraw surfaces.

Add the following code to accomplish this:

```
Sub InitializeDirectX()
 Set DD = DX.DirectDraw4Create("")

 DD.SetCooperativeLevel Form1.hWnd,DDSCL_FULLSCREEN_
 Or DDSCL_EXCLUSIVE

 DD.SetDisplayMode 640, 480, 16, 0, DDSDM_DEFAULT

 DDPrimarySurfDescript.lFlags = DDSD_CAPS Or _
 DDSD_BACKBUFFERCOUNT

 DDPrimarySurfDescript.ddsCaps.lCaps = _
 DDSCAPS_PRIMARYSURFACE Or DDSCAPS_3DDEVICE Or _
 DDSCAPS_COMPLEX Or DDSCAPS_FLIP

 DDPrimarySurfDescript.lBackBufferCount = 1

 Set PrimarySurface = _
 DD.CreateSurface(DDPrimarySurfDescript)
```

```
DDCapsBack.lCaps = DDSCAPS_BACKBUFFER

Set BackSurface = _
PrimarySurface.GetAttachedSurface(DDCapsBack)
```

Once you have initialized DirectDraw, you need to do the same with Direct3D. You also need to determine if you are using a Direct3D hardware card or if not, you should use the Software Renderer. This is actually very easy using an If...Then statement and checking the rmDevice result variable. Lastly, you need to set the quality of the render.

Place the following code in the same procedure:

```
Set RM = DX.Direct3DRMCreate()

BackSurface.SetForeColor RGB(255, 0, 0)

Set rmDevice = _
RM.CreateDeviceFromSurface _
("IID_IDirect3DHALDevice", DD, BackSurface, _
D3DRMDEVICE_DEFAULT)

If rmDevice Is Nothing Then
 Set rmDevice = _
 RM.CreateDeviceFromSurface _
 ("IID_IDirect3DRGBDevice", DD, BackSurface,_
 D3DRMDEVICE_DEFAULT)
End If

If rmDevice Is Nothing Then
 MsgBox "Could not create a Direct3D device"
End If

rmDevice.SetQuality D3DRMLIGHT_ON Or _
D3DRMRENDER_GOURAUD

rmDevice.SetTextureQuality D3DRMTEXTURE_NEAREST
rmDevice.SetRenderMode D3DRMRENDERMODE_BLENDEDTRANSPARENCY
End Sub
```

Creating the Scene

With DirectDraw and Direct3D ready to function, you need to create something to display. The InitializeScene, which if you remember is called after InitializeDirectX in the Form_Load event, is where you create the camera, set the frames, and create the viewport. The camera is being set at X0,Y0,Z-10 so that is centered horizontally and vertically and the Z axis is moved towards you.

The following code completes the tasks:

```
Sub InitializeScene()
 Set rootFrame = RM.CreateFrame(Nothing)
 Set cameraFrame = RM.CreateFrame(rootFrame)
 Set lightFrame = RM.CreateFrame(rootFrame)
 Set objectFrame = RM.CreateFrame(rootFrame)

 rootFrame.SetSceneBackgroundRGB 0, 0, 255

 cameraFrame.SetPosition Nothing, 0, 0, -10
```

Once the viewport has been created, you need to set it to 640x480 resolution. You can also set the position of the light at X5,Y5,Z-25 so that you will see some shadowing. The .x file needs to show up against the blue background, so you can set the light to green.

Here is the code:

```
Set rmViewport = RM.CreateViewport(rmDevice, cameraFrame,_
0, 0, 640, 480)

lightFrame.SetPosition Nothing, 5, 5, -25

Set shadow_light = RM.CreateLightRGB(D3DRMLIGHT_POINT, _
0.5, 0.5, 0.5)

lightFrame.AddLight shadow_light

Set light = RM.CreateLightRGB(D3DRMLIGHT_AMBIENT, 0, 0.5, 0)

rootFrame.AddLight light
```

Next, you need to open the .X file. In this example, you can use the Tiger.X file located in the DirectX SDK. If you are running the application from the CD-ROM rather than typing the code, you will notice that the Tiger.X file has been added to the example directory. This allows you to use the application's path and filename

"Tiger.x". Otherwise, you can hard code the subdirectory into the application. For instance, the Tiger.x file is located at c:\mssdk\samples\multimedia\media\ if you installed the SDK with default settings. You would then replace App.Path with the directory. Lastly, you can set the position of the file at X0,Y0,Z0 so that it is in the center of the scene and then begin the rotation.

This code finishes the Sub procedure:

```
Set meshBuilder = RM.CreateMeshBuilder()
meshBuilder.LoadFromFile App.Path & "\Tiger.x",_
0, 0, Nothing, Nothing
objectFrame.AddVisual meshBuilder
objectFrame.SetPosition Nothing, 0, 0, 0
objectFrame.SetRotation Nothing, 0, 1, 0, 0.25
End Sub
```

THE MAIN LOOP

The RenderLoop Sub procedure is called next. This loop is responsible for continuing to display the contents of the scene until the user decides to end the application.

To begin, you need to set the variable blnIsRunning to True as a way to determine if the loop should continue. Once it is set to False, you should end the program. You need to render the scene and display some basic information to the user about what is happening, how they can exit the program, and how fast the display is being updated.

The following is the code for the main loop.

```
Sub RenderLoop()
 On Local Error Resume Next

 blnIsRunning = True
 lngTick = DX.TickCount()

 Do While blnIsRunning = True
  DoEvents

 rootFrame.Move 0.05

 rmViewport.Clear D3DRMCLEAR_TARGET Or D3DRMCLEAR_ZBUFFER

 rmDevice.Update
```

```
        rmViewport.Render rootFrame

        Call BackSurface.DrawText(5, 5, "# Vertex in Object: "_
        & CStr(meshBuilder.GetVertexCount), False)

        Call BackSurface.DrawText(5, 25, "FPS: " & _
        strFramesText, False)

        Call BackSurface.DrawText(100, 450, _
        "Visual Basic Direct3D Retained Mode _
        - Click the Screen to Exit", False)

        PrimarySurface.Flip Nothing, DDFLIP_WAIT

        intFramesCompleted = intFramesCompleted + 1

        If DX.TickCount >= lngTimeLast + 1000 Then
         lngTimeLast = DX.TickCount
         strFramesText = CStr(intFramesCompleted)
         intFramesCompleted = 0
        End If
       Loop
    End Sub
```

Closing the Application

The program will function at this point in time, but you should not run it. If you were to do so, you would be stuck in an endless loop without a way to escape. You need to provide a simple way for the user to end the application. For the following procedure, you can use the Form_Click event to set the blnIsRunning variable to False, restore the display to the previous settings, and end the application.

You can add this procedure to the application to finish it:

```
    Private Sub Form_Click()
     blnIsRunning = False
     Call DD.RestoreDisplayMode
     Call DD.SetCooperativeLevel(Me.hWnd, DDSCL_NORMAL)
     End
    End Sub
```

When you run the application, you should see something similar to Figure 6.2.

FIGURE 6.2 The running application.

COMPLETE CODE LISTING

The following code is the complete listing for this chapter:

```
Option Explicit

Dim DX As New DirectX7
Dim DD As DirectDraw4
Dim RM As Direct3DRM3

'DirectDraw variables
Dim PrimarySurface As DirectDrawSurface4
Dim BackSurface As DirectDrawSurface4
Dim DDPrimarySurfDescript As DDSURFACEDESC2
Dim DDCapsBack As DDSCAPS2

'RM Device
Dim rmDevice As Direct3DRMDevice3
```

```
'Viewport is what you see on screen
Dim rmViewport As Direct3DRMViewport2

'The Scene
Dim rootFrame As Direct3DRMFrame3
Dim lightFrame As Direct3DRMFrame3
Dim cameraFrame As Direct3DRMFrame3
Dim objectFrame As Direct3DRMFrame3

'Lights and shadow
Dim light As Direct3DRMLight
Dim shadow_light As Direct3DRMLight
Dim shadow As Direct3DRMShadow2

Dim meshBuilder As Direct3DRMMeshBuilder3

Dim lngTimeLast As Long
Dim intFramesCompleted As Integer
Dim strFramesText As String
Dim lngTick As Long

Dim blnIsRunning As Boolean

Private Sub Form_Load()
 InitializeDirectX
 InitializeScene
 Me.Show
 RenderLoop
End Sub

Sub InitializeDirectX()

 Set DD = DX.DirectDraw4Create("")


 DD.SetCooperativeLevel Form1.hWnd, DDSCL_FULLSCREEN Or DDSCL_EXCLUSIVE

 DD.SetDisplayMode 640, 480, 16, 0, DDSDM_DEFAULT

 DDPrimarySurfDescript.lFlags = DDSD_CAPS Or DDSD_BACKBUFFERCOUNT

 DDPrimarySurfDescript.ddsCaps.lCaps = DDSCAPS_PRIMARYSURFACE Or
DDSCAPS_3DDEVICE Or DDSCAPS_COMPLEX Or DDSCAPS_FLIP
```

```
DDPrimarySurfDescript.lBackBufferCount = 1

Set PrimarySurface = DD.CreateSurface(DDPrimarySurfDescript)

DDCapsBack.lCaps = DDSCAPS_BACKBUFFER

Set BackSurface = PrimarySurface.GetAttachedSurface(DDCapsBack)

Set RM = DX.Direct3DRMCreate()

BackSurface.SetForeColor RGB(255, 0, 0)


Set rmDevice = RM.CreateDeviceFromSurface("IID_IDirect3DHALDevice", _
DD, BackSurface, D3DRMDEVICE_DEFAULT)

If rmDevice Is Nothing Then
Set rmDevice = RM.CreateDeviceFromSurface("IID_IDirect3DRGBDevice", _
DD, BackSurface, D3DRMDEVICE_DEFAULT)
End If

If rmDevice Is Nothing Then
MsgBox "Could not create a Direct3D device"
End If

rmDevice.SetQuality D3DRMLIGHT_ON Or D3DRMRENDER_GOURAUD
rmDevice.SetTextureQuality D3DRMTEXTURE_NEAREST
rmDevice.SetRenderMode D3DRMRENDERMODE_BLENDEDTRANSPARENCY

End Sub

Sub InitializeScene()
 Set rootFrame = RM.CreateFrame(Nothing)
 Set cameraFrame = RM.CreateFrame(rootFrame)
 Set lightFrame = RM.CreateFrame(rootFrame)
 Set objectFrame = RM.CreateFrame(rootFrame)

 rootFrame.SetSceneBackgroundRGB 0, 0, 255

 cameraFrame.SetPosition Nothing, 0, 0, -10

 Set rmViewport = RM.CreateViewport(rmDevice, cameraFrame, 0, 0, 640, _
 480)
```

```
  lightFrame.SetPosition Nothing, 5, 5, -25

  Set shadow_light = RM.CreateLightRGB(D3DRMLIGHT_POINT, 0.5, 0.5, 0.5)

  lightFrame.AddLight shadow_light

  Set light = RM.CreateLightRGB(D3DRMLIGHT_AMBIENT, 0, 0.5, 0)
  rootFrame.AddLight light

  Set meshBuilder = RM.CreateMeshBuilder()
  meshBuilder.LoadFromFile App.Path & "\Tiger.x", 0, 0, Nothing, Nothing

  objectFrame.AddVisual meshBuilder

  objectFrame.SetPosition Nothing, 0, 0, 0

  objectFrame.SetRotation Nothing, 0, 1, 0, 0.25

End Sub

Sub RenderLoop()
 On Local Error Resume Next

  blnIsRunning = True
  lngTick = DX.TickCount()

  Do While blnIsRunning = True
  DoEvents

  rootFrame.Move 0.05

  rmViewport.Clear D3DRMCLEAR_TARGET Or D3DRMCLEAR_ZBUFFER

  rmDevice.Update

  rmViewport.Render rootFrame

  Call BackSurface.DrawText(5, 5, "# Vertex in Object: " &
CStr(meshBuilder.GetVertexCount), False)

  Call BackSurface.DrawText(5, 25, "FPS: " & strFramesText, False)

  Call BackSurface.DrawText(100, 450, "Visual Basic Direct3D Retained
Mode – Click the Screen to Exit", False)
```

```
    PrimarySurface.Flip Nothing, DDFLIP_WAIT

    intFramesCompleted = intFramesCompleted + 1

    If DX.TickCount >= lngTimeLast + 1000 Then
    lngTimeLast = DX.TickCount
    strFramesText = CStr(intFramesCompleted)
    intFramesCompleted = 0
    End If
    Loop
End Sub

Private Sub Form_Click()
 blnIsRunning = False
 Call DD.RestoreDisplayMode
 Call DD.SetCooperativeLevel(Me.hWnd, DDSCL_NORMAL)
 End
End Sub
```

CHAPTER REVIEW

In this chapter, you learned a great deal about DirectX, and specifically Direct-Draw and Direct3D. If you plan to use Direct3D in a large application or game, you will definitely want to invest in the time to learn the Immediate Mode functionality. The Help files included in the DirectX SDK are invaluable as a learning reference. It's the only mode that Microsoft is directly supporting in version 8 and, in all probability, the lack of support for Retained Mode will continue in later versions. That said, Retained Mode is a great way to learn the basics of Direct3D and it provides a quick and relatively painless solution when it is used in a small application.

You could easily modify this application to create a program that would display any DirectX Model.

7 Visual Basic Screen Saver

A Windows screen saver program is nothing more than a standard executable file (.exe) renamed with the extension .scr and saved in the Windows directory. In theory, you could simply rename any program with an .scr extension and it would work correctly. However, in order to function correctly, there are a few additional requirements. First, it should stop execution when a mouse is moved or clicked, or when a certain key is pressed. It should also launch only one instance of itself. After all, you don't want 25 screen savers loaded into memory at a single time.

PROJECT OVERVIEW

In this chapter, you will develop a Windows Screen Saver that draws a gradient background and displays a text message at random locations. The text information moves every second while the background will also change every five seconds. A sample of the screen is located in Figure 7.1.

FIGURE 7.1 A Visual Basic Screen Saver in action.

Beginning the Project

Setting Up the Main Form

The first step in the project is to set up the main form. Begin by creating a new VB project and setting some properties for the default main form. This form needs to resize itself to fill the entire screen, and it should not have a title bar or control buttons at the top.

Set the following properties for the form:

- AutoRedraw: True

- BorderStyle: 0-None

- ControlBox: False

- MaxButton: False

- MinButton: False

- WindowState: 2-Maximize

Once the properties are set, you need to add a few controls. If you look back at Figure 7.1, the final screen saver displays a gradient background and some text information. A label control will be used to store the text information, so add it to the form. During the project, you will add some code to move the label around the form to random locations, so its position isn't important at this time. Next, add two timer controls to the form. Again, positioning isn't important, as they are not displayed at runtime.

The Label Control Properties

The Label control you placed on the form in the previous step will need to have some properties changed. You could programmatically change these or in our case, you can simply change their properties using the Properties window:

- Name: lblBanner

- Autosize: True

- BackStyle: 0-Transparent

- Caption: My Screen Saver

You can also change the font to something that is a little larger and easier to read. You do this by clicking the ellipsis(...) button that is located in the Font property of the Properties window. You'll already be familiar with the Font Selection window, as it is identical to other Windows programs. Select a large font size such as 18 or 24.

The Timer Properties

The timer controls need only a single change to their properties, the Interval property. Set the Timer1 control to an interval of 5000 milliseconds (5 seconds) and the Timer2 control to an interval of 1000 milliseconds (1 second). The first timer will execute the code in the Timer1_Timer event every 5 seconds. In this event, we'll place code for changing the gradient. Timer2 will randomly move lblBanner every second.

PROGRAM DECLARATIONS

The following variable declarations need to be made in the Code window of Form1:

```
Dim Red As Integer
Dim Green As Integer
Dim Blue As Integer
Dim SquareHeight As Integer
Dim DrawLine As Integer
Dim SquareTop As Integer
Dim SquareBottom As Integer
Dim XPosition As Integer
Dim YPosition As Integer
Dim CountMouse As Integer
```

You'll notice that all the variables are type Integer. You can alternately set the variables like this:

```
Dim Red, Green, Blue As Integer
```

This saves some space and time if you wish to do so.

FORM_LOAD

The Form_Load event is called when the application is first executed. In this event, you need to handle several special situations for executing a screen saver. The parameters are passed by Windows to allow certain functionality to be executed. For instance, if you have a configuration form for a screen saver like Figure 7.2, Windows passes "/c" when it wants to configure the screen saver. The following list details these parameters:

- "/P": Preview Mode for the screen saver. For this to work, you would have to handle a great deal of code and declarations. We'll just ignore this call.

- "/A": The Password dialog box (same as a dialog window) is called with this command. The screen saver you are building is not password protected, and is not used or discussed in this tutorial, but it isn't hard to figure out on your own if you're interested.

- "/C": The Configuration window is what you see when you click the "Settings" button in the Screen Saver Tab of Desktop Properties. This screen saver will not have any configuration, but it would be easy to change it to allow the user to enter information to be stored by the screen saver. One such idea would be to allow the screen saver to have user selectable text information to be displayed. Another idea would be to change the timers to slow down or speed up the random movement of the label.

- "/S": Runs the screen saver.

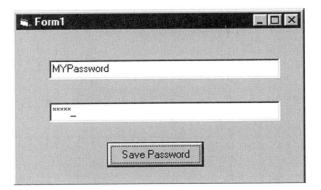

FIGURE 7.2 Configuration options in a screen saver.

The following code takes care of the Form Load event:

```
Private Sub Form_Load()
  If LCase(Command$) = "/c" Then
  MsgBox "This screen saver does not provide setup options."
  Unload Me
  Exit Sub
  End If
```

```
    If LCase(Command$) <> "/s" Or App.PrevInstance Then
      Unload Me
      Exit Sub
    End If
    Randomize
End Sub
```

This Form_Load event checks to see what extra parameters have been set and it also initializes the Visual Basic random numbers to make sure that what we get later from the random functions will truly be random.

Once the form is loaded, the Form_Activate event is executed. We use this to call our RandomColors Sub procedure rather than the Form Load event. If we did this in the Form Load event, only a small section of the form would be drawn to.

That's the only call for the procedure. The Timer1_Timer event will also need to execute this code to redraw the screen every five seconds. You can enter the following code below the previous code:

```
Private Sub Form_Activate()
  RandomColors
End Sub

Private Sub Timer1_Timer()
  RandomColors
End Sub
```

GENERATE RANDOM COLORS

The RandomColors procedure that Timer1 and Form_Activate call will use a series of random number functions to select a random red, green and blue integer value. These values will then be used by the GradientForm procedure to actually draw the screen.

The syntax for the Rnd function is as follows: Rnd(Number). The number returned will be between 0 and 1. To get the number into the required values, you can simply multiply a value by the result and use the Int function to remove any decimals.

The following code sets the red, green, and blue values and then calls the GradientForm procedure:

```
Public Sub RandomColors()
  Red = Int(Rnd * 256)
```

```
    Green = Int(Rnd * 256)
    Blue = Int(Rnd * 256)
    GradientForm
End Sub
```

DRAWING GRADIENTS

The GradientForm procedure contains the most interesting code for this project. Basically, the procedure will draw a number of increasingly darkening squares down the form from top to bottom. The procedure will begin by taking the form height and divide it by 100, storing the value in the variable SquareHeight. The number 100 isn't any magic number, you could use any number you would like. It's just an easy number to use for calculations so that you can watch what is happening.

Next, you set the SquareTop variable to 0 so that it draws the first square at 0. Next, a variable named SquareBottom is given a value equal to the SquareTop +SquareHeight. In other words, the bottom is one hundredth of the form's height.

Next, a For...Next loop is started with the variable DrawLine. The loop goes from 1 to 100 (If you change the 100 in the previous step, you also need to change it here to reflect the value). Inside the loop, a square is drawn and the colors are adjusted to become darker by subtracting an integer value of 2 from them. If you'd like, you can adjust these value as well. Lastly, the procedure needs to set the SquareTop variable equal to the position of the previous Square Bottom and then add one hundredth of the form's height to the SquareTop value.

```
Private Sub GradientForm()
  SquareHeight = (Me.Height / 100)
  SquareTop = 0
  SquareBottom = SquareTop + SquareHeight

  For DrawLine = 1 To 100

    Me.Line (0, SquareTop)-(Me.Width, SquareBottom), _
    RGB(Red, Green, Blue), BF

    Red = Red - 2
    Green = Green - 2
    Blue = Blue - 2
    If Red <= 0 Then Red = 0
    If Green <= 0 Then Green = 0
    If Blue <= 0 Then Blue = 0
```

```
      SquareTop = SquareBottom
      SquareBottom = SquareTop + SquareHeight
   Next DrawLine
End Sub
```

THE SECOND TIMER

The screen saver is almost finished. You have only a few minor items to left to do. First, you need to create the code to move lblBanner around the screen. If you remember, this is supposed to take place when the Timer2_Timer event occurs. You can use the Rnd function again in this procedure, using two variables XPosition and YPosition and setting them equal to a random value. You can also set the lblBanner's ForeColor property to a random RGB value and then use the Move method to move it. The Move method will set the top and left position of the label to an X,Y value on the screen (or off). You need to create the random functions so that the X and Y positions will be on screen.

Here is the code for the procedure:

```
Private Sub Timer2_Timer()
  XPosition = Int(Rnd * (ScaleWidth - lblBanner.Width))
  YPosition = Int(Rnd * (ScaleHeight - lblBanner.Height))
  lblBanner.ForeColor = RGB(Int(Rnd * 256), Int(Rnd * 256), Int(Rnd * 256))
  lblBanner.Move XPosition, YPosition
End Sub
```

ENDING THE SCREEN SAVER

The last programming step involves setting up situations that will stop the screen saver from executing. If you left this step out, the screen saver would be locked in an endless loop. You can use the Form_Click event, the Form_KeyDown event and the Form_MouseMove event to end the application if any of these situations occur. The first two simply use the Unload statement to end the application.

Here is their code:

```
Private Sub Form_Click()
  Unload Me
End Sub
```

```
Private Sub Form_KeyDown(KeyCode As Integer, Shift As Integer)
 Unload Me
End Sub
```

The MouseMove event is a little more difficult. When the form is maximized, Windows triggers the MouseMove event. This would cause an unnecessary ending to the program every time it loaded up. To correct this, you can use the variable CountMouse, which will simply be used as a counter. Once the counter equals a certain value, the screen saver can exit. This is a very simple solution to the problem.

Here is the code for the last procedure:

```
Private Sub Form_MouseMove(Button As Integer, Shift As Integer,_
XPosition As Single, YPosition As Single)

  If CountMouse < 5 Then
    CountMouse = CountMouse + 1
  Else
    Unload Me
  End If
End Sub
```

TESTING THE PROJECT IN THE IDE

If you were to run the application in the IDE, it would simply stop executing almost immediately. This is because of the Form_Load event, which checks to see if a screen saver value is passed to it. Without the "/S" command line being passed, the program will simply exit.

You could handle this in several ways. First, you could simply add a single apostrophe (') before the lines that check for this in the Form_Load event. (Visual Basic ignores all lines beginning with this character. These lines are called comments.) This would work, but it's not a good idea to get into the habit of changing a program so that it is forced to run under certain conditions. With a simple application like this, it's not a problem, but with very large programs, you need to make sure the entire application is functioning correctly. Commenting out the lines would not allow you to do this.

You could also add another check to the If...Then statement. You could alter it to read something like the following: If LCase(Command$) = "/s" Or App.PrevInstance or Command$= "" Then ...

This would effectively handle the problem of not running in the IDE, but again, it's not a perfect solution. You would also need to include additional checks for the other parameters that were previously handled with the not equal (<>) in the If...Then statement.

The best solution would be to simply add the command line parameter to the Visual Basic IDE which does not alter the application in any way. In the Project menu, you will find a Project Properties menu item. Select it, then in the Make Tab, you can enter the /S command under Command Line Arguments. Your screen should like something like Figure 7.3. Click the OK button and you are ready to test the program.

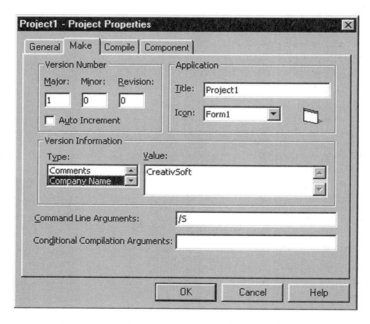

FIGURE 7.3 Command line arguments need to be addressed in the IDE.

COMPILING AND USING THE SCREEN SAVER

Once you have saved the project and tested its execution, you need to make a SCR file from it. Actually, you need to create a standard EXE file and then rename it to SCR. The Make EXE option is located in the File menu. You need to compile the application to the Windows directory (it's usually c:\windows) and then rename the file to something like VBSaver.SCR. If you've done everything correctly, the screen saver should be listed in the available options in the Control Panel Display applet.

If you run the screen saver, you will notice that the mouse pointer remains visible. If you'd like, you could add a little more code to do this. The first item is a call to the Windows API, and should be added to the top of the General Declarations section for Form1:

```
Private Declare Function ShowCursor& Lib "user32" (ByVal bShow As Long)
```

Next, you can create a new procedure called EndIt, and inside it place the following code:

```
Private Sub EndIt()
 ShowCursor True
 Unload Me
End Sub
```

Replace all of the Unload Me statements (besides the one located in this Sub procedure) with calls to EndIt. This will send a True value to the API function, which will turn the mouse pointer on when the project exits. To turn it off, simply place the following line at the end of the GradientForm Sub procedure after the Next DrawLine statement:

```
ShowCursor False
```

This will safely turn the mouse pointer off if the program executes a gradient fill. Alternately, you could create an icon that was completely transparent and load this icon whenever the program runs, returning to a standard icon when it was finished.

COMPLETE CODE LISTING

The following code is the complete listing for this chapter:

```
Option Explicit
Private Declare Function ShowCursor& Lib "user32" (ByVal bShow As Long)

Dim Red As Integer
Dim Green As Integer
Dim Blue As Integer
Dim SquareHeight As Integer
```

```
Dim DrawLine As Integer
Dim SquareTop As Integer
Dim SquareBottom As Integer
Dim XPosition As Integer
Dim YPosition As Integer
Dim CountMouse As Integer

Private Sub Form_Load()

 If LCase(Command$) = "/c" Then
  MsgBox "This screen saver does not provide setup options."
  EndIt
  Exit Sub
 End If

 If LCase(Command$) <> "/s" Or App.PrevInstance Then
  EndIt
  Exit Sub
 End If

 Randomize
End Sub

Private Sub Form_Activate()
 RandomColors
End Sub

Private Sub Timer1_Timer()
 RandomColors
End Sub

Public Sub RandomColors()

 Red = Int(Rnd * 256)
 Green = Int(Rnd * 256)
 Blue = Int(Rnd * 256)

 GradientForm
End Sub

Private Sub GradientForm()
 SquareHeight = (Me.Height / 100)
 SquareTop = 0
 SquareBottom = SquareTop + SquareHeight
```

```
For DrawLine = 1 To 100

 Me.Line (0, SquareTop)-(Me.Width, SquareBottom), RGB(Red, Green,
Blue), BF

  Red = Red - 2
  Green = Green - 2
  Blue = Blue - 2

  If Red <= 0 Then Red = 0
  If Green <= 0 Then Green = 0
  If Blue <= 0 Then Blue = 0

  SquareTop = SquareBottom
  SquareBottom = SquareTop + SquareHeight
 Next DrawLine
 ShowCursor False
End Sub

Private Sub Timer2_Timer()

 XPosition = Int(Rnd * (ScaleWidth - lblBanner.Width))
 YPosition = Int(Rnd * (ScaleHeight - lblBanner.Height))

 lblBanner.ForeColor = RGB(Int(Rnd * 256), Int(Rnd * 256), Int(Rnd *
256))
 lblBanner.Move XPosition, YPosition
End Sub

Private Sub Form_Click()
 EndIt
End Sub

Private Sub Form_KeyDown(KeyCode As Integer, Shift As Integer)
 EndIt
End Sub

Private Sub Form_MouseMove(Button As Integer, Shift As Integer,
XPosition As Single, YPosition As Single)

 If CountMouse < 5 Then
  CountMouse = CountMouse + 1
 Else
  EndIt
```

```
    End If
End Sub

Private Sub EndIt()
 ShowCursor True
 Unload Me
End Sub
```

CHAPTER REVIEW

In this chapter, you learned about several new items and touched upon many we have already looked at. Among the new items were random numbers and the Randomize statement and the difference between Form_Load and Form_Activate. You also learned about the minor difference between a screen saver and a standard executable file and the Command Line options that need to be checked for a screen saver, and how to deal with them in the Visual Basic IDE.

8 Design a Word Processor

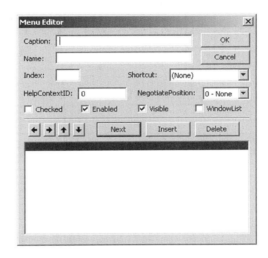

In this chapter, we'll build a Visual Basic Word Processor. It's not going to be a replacement for Microsoft Word, but it will have enough features to make it useful. While you are developing the VB Word Processor, you'll be exposed to a variety of issues including menus and the Rich Textbox control.

THE PROJECT

Overview

A word processor should have a set of menus on a menu bar along the top of the page that offers access to the most popular features. The top-level menu items (those with their names on the menu bar) should consist of entries such as File and

Edit, which will be clicked on to reveal a selection of other functions that pertain to the top-level item. For example, if you wish to open or save your document, you can find a function to do this under the File menu item. Likewise, you can find options for copying and pasting under Edit.

Introducing the Rich Textbox and Common Dialog Controls

You could use the intrinsic Text Box Control for developing this project, but you would be limited in functionality. Instead, Microsoft provides the Rich Textbox ActiveX control with Visual Basic. This control can open and save documents that conform to the Microsoft established Rich Text Format (RTF) standard for specifying formatting of documents. The files are simply ASCII files with special commands to indicate formatting information, such as fonts and margins.

To begin, open the Components window by pressing CTRL-T or right-clicking on the Toolbox and selecting Components from the pop-up menu. Next, select Microsoft Rich Textbox Control and the Microsoft Common Dialog Control from the list of components in the window. Once you select them, you can click OK to close the window. At this point, you should see the two controls available in the Toolbox. Place both controls on the form in an arrangement similar to Figure 8.1. For this form, the dimensions are approximately 7080 width and 6070 for the height while the Rich Textbox is about 6700 for the width and 4600 for its height.

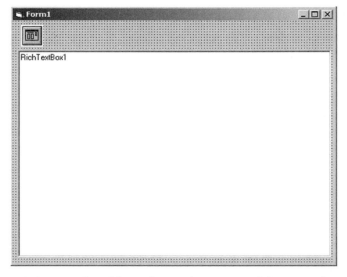

FIGURE 8.1 The Rich Textbox and Common Dialog controls are the only controls needed on the form.

In the previous step you placed a Common Dialog control along with the Rich Textbox control. The Common Dialog is a control that allows you to use standard Windows dialog boxes such as Open, Save and Color to name a few. It's an ActiveX control provided freely by Microsoft and is the easiest way to add dialog boxes to your application. Before we begin writing code, you first need to add a menu to the application.

DESIGNING A MENU

Menus are designed to be a fairly intuitive way to access certain functions in a program. It's important to think carefully about where you should place the commands. For instance, you wouldn't usually find an Open command placed under the Edit top-level menu item. This program will have four top-level menu items: File, Edit, Font and Alignment. Beneath these levels will be additional commands that correspond to the top-level. Although it's not a requirement, you should try to follow the standard conventions used in other Windows applications. The File menu should be placed at the left-hand side of the window followed by the Edit menu and so on.

With these thoughts in mind, here is the menu structure for this project:

File	*Edit*	*Fonts*	*Alaignment*
New	Copy	Font	Center
Open	Cut	Uppercase	Left
Save	Paste	Bold	Right
Print	BackColor	Italic	
Exit		Normal	

Adding the menu items is very easy in Visual Basic. First, you open the form that you need to place the menu items on. Next, click the Menu Editor toolbar button. The Menu Editor will appear and should look something like Figure 8.2.

You'll notice in Figure 8.2 that there are a number of fields available for each menu entry. Of these fields, the Caption and Name properties are the only required entries. The three letters "mnu" are commonly used as a prefix for the name of an object. For instance, the File entry could use mnuFile as a name. For the sample project, I simply set the names and captions equal to one another without any "mnu" prefix.

If you remember the menu arrangement, the first listing we need is File. You can enter &File as the Caption and File as the Name. The "&" instructs VB that you wish to have an underline beneath the letter immediately following it. Therefore,

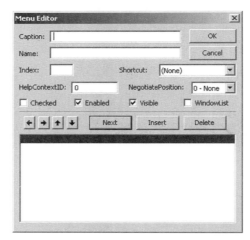

FIGURE 8.2 The VB Menu Editor allows menu creation to be an easy process.

instead of File you'll actually get File. The underline is significant as it allows you to use the Alt key followed by the letter F to select it.

The next step is to click on the Next button, which will move you to the next line. Click on the right arrow button, and you will see four dots appear. It is this process that allows you to place items beneath the top-level menu entries. The New menu item is next, so we can use New as a Caption and New as the Name. At this time, click on the Next button and we'll create the Open menu option.

The Open option will be constructed the same way as the previous entry with a single addition — an entry for the shortcut combination CTRL-O. This allows the Open command to be executed when the end user of the program presses the key combination.

With this information, you can continue on with the rest of the menu. After finishing the File menu with the Exit entry, you can simply create a new menu item by clicking the left arrow to make the Edit entry a top-level menu item. One last item that is different from the rest is the use of a hyphen ("-") in the Caption box. If you enter a hyphen as a menu caption, you will place a divider in the box. The divider can be used to set apart options beneath the top-level menu item. For instance, in our example, you can place a divider after Print and before Exit to separate the options.

Design a Word Processor **131**

The following list details the final Menu Editor along with any shortcut options. (If you have already added all the items for the File menu, after adding the last item, go back up to the New item and add the keyboard shortcut). When you finish, click OK to return to the Form Designer Window.

Menu Caption	Name	Shortcut
&File		
File	New	CTRL-N
Open	Open	CTRL-O
Save	Save	CTRL-S
Print	Print	CTRL-P
-	Divider	
Exit	Exit	CTRL-Q
&Edit		
Copy	Copy	CTRL-C
Cut	Cut	CTRL-X
Paste	Paste	CTRL-V
BackColor	Backcolor	
&Fonts		
Font	Font	CTRL-F
Uppercase	Uppercase	CTRL-U
Bold	Bold	CTRL-B
Italic	Italic	CTRL-I
Normal	Normal	CTRL-T
&Alignment		
Center	Center	
Left	Left	
Right	Right	

ADDING SOME CODE

File Menu

Now that we have a completed menu bar with menu options, it's time to add code to each item that we created. It's easiest to start with the File menu since it's first. Visual Basic provides an easy way to get to the events that are raised by the menu items. You can simply click on the menu item on the form while you are in Design Mode. This will open the Code window and place your cursor at the appropriate location. For instance, if you click the File menu item, then the New item, you will see a screen similar to Figure 8.3.

FIGURE 8.3 The Code window is opened automatically when a menu item is clicked at design time.

Once you are inside the Code window in the New Click_Event() Sub procedure, you can add code which will use a message box to make sure that you really want to create a new file, and if so, will set the RichTextBox control to an empty string. Otherwise, it will exit the procedure leaving the RichTextBox untouched. Here is the code:

```
Private Sub New_Click()
Dim Result As String
 Result = MsgBox("Close this file?", vbYesNo, _
 "VB Word Processor")
 If Result = vbNo Then Exit Sub
 RichTextBox1 = ""
End Sub
```

While the previous code is very similar to that of previous chapters, the next section will introduce you to the Common Dialog control that you placed in the Toolbox at the beginning of the chapter.

Common Dialog Control

To use the Common Dialog control, you have a few properties and methods you must set first. The following list details some of the more common properties and methods for the Common Dialog control:

Property/Method	Used For
Filter	File Filter
Filename	Selected Filename
ShowOpen	Display Open
ShowSave	Display Save As
ShowColor	Display Color
ShowFont	Display Font
ShowPrinter	Display Print
ShowHelp	Invokes Windows Help

The first menu item that uses the Common Dialog control is File | Open. The Open_Click event needs to display an Open dialog box and will use several of the properties and methods we mentioned to set a filter for specific files and set the Rich Textbox equal to the filename of the Common Dialog box. Lastly, we'll set the caption of the form to display the currently open file.

Before you enter the code, it's important to understand a little about the syntax of the Common Dialog filter, which will limit the types of files that are displayed. For our word processor, we need to limit the files to text and richtext, along with an option to display all files. You can enter the following code for the Open_Click event:

```
Private Sub Open_Click()
  CommonDialog1.Filter = "All Files (*.*)|*.*|Text Files" & _
  "(*.txt)|*.txt|Rich Text Files (*.rtf)|*.rtf"
  CommonDialog1.ShowOpen
  RichTextBox1.FileName = CommonDialog1.FileName
  Form1.Caption = "VB Word Processor - " & _
  CommonDialog1.FileName
End Sub
```

Saving files works very similar to opening them. You can enter the following code:

```
Private Sub Save_Click()
 CommonDialog1.Filter = "Rich Text Files (*.rtf)|*.rtf"
 CommonDialog1.ShowSave
 RichTextBox1.SaveFile CommonDialog1.FileName
 Form1.Caption = "VB Word Processor - " & _
 CommonDialog1.FileName
End Sub
```

Printing

The Common Dialog displays printer options as well. Fortunately, the Common Dialog is very simple to use in this capacity as well. The first step is to set the default printer property of the Common Dialog equal to True. We need a way to exit the subroutine if a user clicks on the Cancel button of the Print Dialog box. This can be accomplished by setting the Common Dialog's CancelError equal to True. Once that is set, whenever the user clicks Cancel, an error is created that can be used to exit the subroutine without printing.

There are only a few additional steps to complete this part of the application. First, we need to determine if there is any data in the Rich Textbox (you probably wouldn't want to print a page with nothing on it). Once we check for data, we need to set some flags that instruct the Common Dialog that there will not be any page numbers and finally print the page using the printer handle. The following code is the entire Sub procedure needed for printing the contents of the Rich Textbox:

```
Private Sub Print_Click()
 CommonDialog1.PrinterDefault = True
 CommonDialog1.CancelError = True
 CommonDialog1.Flags = cdlPDReturnDC + cdlPDNoPageNums
 If RichTextBox1.SelLength = 0 Then
 CommonDialog1.Flags = CommonDialog1.Flags + cdlPDAllPages
 Else
 CommonDialog1.Flags = CommonDialog1.Flags + _
 cdlPDSelection
 End If
 On Error Resume Next
 CommonDialog1.ShowPrinter
 If Err Then
 Exit Sub
 End If
```

```
    RichTextBox1.SelPrint (Printer.hDC)
End Sub
```

The last menu item located beneath the top-level File menu is Exit. This is the easiest of all the Sub procedures in this project as it's only a single line:

```
Private Sub Exit_Click()
 Unload Me
End Sub
```

Cut, Copy, and Paste

The File menu is now completed and we can move on to the Edit menu, which contains the Cut, Copy, Paste, and Background Color options. The Cut and Copy commands that are used in this sample are very simple and utilize only standard VB code and operations. For the Paste command, we need to use an API function that will allow us to paste not only text information but also pictures as well. Without the API call, we would only have access to copying and pasting text information.

To begin, let's set up the API declarations and information. We'll be using the SendMessage API call along with a constant WM Paste. The following code can be placed in the General Declarations section of the Code window:

```
Option Explicit
Private Declare Function SendMessage Lib "user32" Alias _ "SendMessageA"
(ByVal hwnd As Long, _
ByVal wMsg As Long, ByVal wParam As Long, lParam As Any) As Long

Private Const WM_PASTE = &H302
```

With the API declaration out of the way, the rest of the code is very simple. The Copy and Cut commands use the VB Clipboard object. The Clipboard object can contain several pieces of data as long as each piece is in a different format. For example, you can use the SetData method to put a bitmap on the Clipboard with the vbCFDIB format, and then use the SetText method with the vbCFText format to put text on the Clipboard. You can then use the GetText method to retrieve the text or the GetData method to retrieve the graphic.

Data on the Clipboard is lost when another set of data is placed on the Clipboard, either through code or a menu command. Unfortunately, we cannot paste bitmaps into the Rich Textbox without the use of the API. Therefore, here is the necessary code for the Cut, Copy, and Paste commands:

```
Private Sub Copy_Click()
 Clipboard.Clear
```

```
Clipboard.SetText RichTextBox1.SelText, vbCFRTF

End Sub
Private Sub Cut_Click()
 Clipboard.Clear
 Clipboard.SetText RichTextBox1.SelText, vbCFRTF
 RichTextBox1.SelText = ""
End Sub

Private Sub Paste_Click()
 'Use instead of Clipboard.GetText(vbCFRTF)
 SendMessage RichTextBox1.hwnd, WM_PASTE, 0, 0
End Sub
```

The last menu item located beneath the top-level Edit menu is Background Color. To change the background of the Rich Textbox by using the BackColor property, you can utilize the Common Dialog control and its ShowColor method. The following code is all that is needed:

```
Private Sub BackColor_Click()
 On Error GoTo ErrorHandler
 CommonDialog1.ShowColor
 RichTextBox1.BackColor = CommonDialog1.Color
ErrorHandler:
End Sub
```

The Font Menu

The Font menu is the next top-level menu item that needs to be addressed and is the easiest menu that we've dealt with. It sets a few properties and methods of the Common Dialog and Rich Textbox, which is very similar to what we've already done. There are a few new items that will be introduced. First, the Common Dialog Control Flags property must be set before the ShowFont method is called. You have the option of displaying printer fonts (cdlCFPrinterFonts), screen fonts (cdlCFScreenFonts), or both printer and screen fonts (cdlCFBoth). For this application, we'll display both types of fonts.

The following code will display the Common Dialog control with both printer and screen fonts and will set the Rich Textbox control's font to the Common Dialog selected font:

```
Private Sub Font_Click()
 On Error GoTo ErrorHandler
```

```
    CommonDialog1.Flags = cdlCFBoth
    CommonDialog1.ShowFont
    RichTextBox1.Font = CommonDialog1.FontName
ErrorHandler:
End Sub
```

To change the properties of a font in the Rich Textbox control, you can use the built in properties that are available. There are properties available for Bold, Italic and Uppercase, but there does not exist a single property to set the font back to Normal. As a result, you can simply set each of the other properties to False to regain a Normal font.

You may notice that we're using the uppercase property in this example. There are additional properties (like underline) that can be used alongside those that we have chosen. Uppercase was chosen so that you would be exposed to a totally different type of property that dealt with additional code.

The following code sets all of the font attributes:

```
Private Sub Uppercase_Click()
 RichTextBox1.SelText = UCase(RichTextBox1.SelText)
End Sub

Private Sub Bold_Click()
 RichTextBox1.SelBold = Not RichTextBox1.SelBold
End Sub

Private Sub Italic_Click()
 RichTextBox1.SelItalic = Not RichTextBox1.SelItalic
End Sub

Private Sub Normal_Click()
 RichTextBox1.SelBold = False
 RichTextBox1.SelItalic = False
 RichTextBox1.SelUnderline = False
End Sub
```

The final font attributes for alignment are set much the same. The following code handles the entire Alignment menu:

```
Private Sub Center_Click()
 RichTextBox1.SelAlignment = 2 'Center
End Sub
```

```
Private Sub Left_Click()
 RichTextBox1.SelAlignment = 0 'Left
End Sub

Private Sub Right_Click()
 RichTextBox1.SelAlignment = 1 'Right
End Sub
```

The Final Steps

To finish the project, we have a few additional steps. First, we need to move and resize the Rich Textbox control to the width and height of the form as it is resized. The width needs to be set equal to the width of the form – 75 (so we can display a scroll bar) while the height is set equal to the height of the Form. The following procedures are the final code:

```
Private Sub Form_Load()
 RichTextBox1.Move 0, 0, Form1.Width, Form1.Height
 RichTextBox1.Text = ""
 Form1.Caption = "VB Word Processor"
End Sub

Private Sub Form_Resize()
 RichTextBox1.Width = Form1.Width - 75
 RichTextBox1.Height = Form1.Height
End Sub
```

COMPLETE CODE LISTING

The following code is the complete code listing:

```
Option Explicit

Private Declare Function SendMessage Lib "user32" Alias "SendMessageA"
(ByVal hwnd As Long, _
ByVal wMsg As Long, ByVal wParam As Long, lParam As Any) As Long

Private Const WM_PASTE = &H302
```

```vb
Private Sub Form_Load()
 RichTextBox1.Move 0, 0, Form1.Width, Form1.Height
 RichTextBox1.Text = ""
 Form1.Caption = "VB Word Processor"
End Sub


Private Sub BackColor_Click()
 'If an error occurs goto Errorhandler
 On Error GoTo ErrorHandler

 CommonDialog1.ShowColor
 RichTextBox1.BackColor = CommonDialog1.Color

ErrorHandler:
End Sub


Private Sub Bold_Click()
 RichTextBox1.SelBold = Not RichTextBox1.SelBold
End Sub


Private Sub Center_Click()
 RichTextBox1.SelAlignment = 2 'Center
End Sub


Private Sub Copy_Click()
 Clipboard.Clear
 Clipboard.SetText RichTextBox1.SelText, vbCFRTF
End Sub


Private Sub Cut_Click()
 Clipboard.Clear
 Clipboard.SetText RichTextBox1.SelText, vbCFRTF
 RichTextBox1.SelText = ""
End Sub


Private Sub Exit_Click()
 Unload Me
End Sub
```

```vb
Private Sub Font_Click()
'If an error occurs goto Errorhandler
 On Error GoTo ErrorHandler
' Flags property must be set
 ' to cdlCFBoth,
 ' cdlCFPrinterFonts,
 ' or cdlCFScreenFonts before
 ' using ShowFont method
 CommonDialog1.Flags = cdlCFBoth
 CommonDialog1.ShowFont
 RichTextBox1.Font = CommonDialog1.FontName

'If an error occurs when Cancel is clicked
ErrorHandler:
End Sub


Private Sub Form_Resize()
 RichTextBox1.Width = Form1.Width - 75
 RichTextBox1.Height = Form1.Height
End Sub


Private Sub Italic_Click()
 RichTextBox1.SelItalic = Not RichTextBox1.SelItalic
End Sub


Private Sub Left_Click()
 RichTextBox1.SelAlignment = 0 'Left
End Sub


Private Sub New_Click()
Dim Result As String
 Result = MsgBox("Close this file?", vbYesNo, "VB Word Processor")
 If Result = vbNo Then Exit Sub
 RichTextBox1 = ""
End Sub


Private Sub Normal_Click()
```

```
  RichTextBox1.SelBold = False
  RichTextBox1.SelItalic = False
  RichTextBox1.SelUnderline = False
End Sub


Private Sub Open_Click()
  CommonDialog1.Filter = "All Files (*.*)|*.*|Text Files" & _
  "(*.txt)|*.txt|Rich Text Files (*.rtf)|*.rtf"
  CommonDialog1.ShowOpen
  RichTextBox1.FileName = CommonDialog1.FileName
  Form1.Caption = "VB Word Processor - " & CommonDialog1.FileName
End Sub

Private Sub Paste_Click()
  'Use instead of Clipboard.GetText(vbCFRTF)
  SendMessage RichTextBox1.hwnd, WM_PASTE, 0, 0
End Sub


Private Sub Print_Click()
  'Print Dialog / Return Printer
  CommonDialog1.PrinterDefault = True
  CommonDialog1.CancelError = True

  'Set flags - no page numbers, return the selected printer
  CommonDialog1.Flags = cdlPDReturnDC + cdlPDNoPageNums
  If RichTextBox1.SelLength = 0 Then
  CommonDialog1.Flags = CommonDialog1.Flags + cdlPDAllPages
  Else
  CommonDialog1.Flags = CommonDialog1.Flags + cdlPDSelection
  End If

  'Enables error for cancel error
  On Error Resume Next
  'Display the print dialog box
  CommonDialog1.ShowPrinter

  If Err Then
  'If error then exit
  Exit Sub
  End If

  'Prints RichTextBox
```

```
    RichTextBox1.SelPrint (Printer.hDC)

End Sub


Private Sub Right_Click()
 RichTextBox1.SelAlignment = 1 'Right
End Sub


Private Sub Save_Click()
 CommonDialog1.Filter = "Rich Text Files (*.rtf)|*.rtf"
 CommonDialog1.ShowSave

 RichTextBox1.SaveFile CommonDialog1.FileName
 Form1.Caption = "VB Word Processor - " & CommonDialog1.FileName
End Sub


Private Sub Uppercase_Click()
 RichTextBox1.SelText = UCase(RichTextBox1.SelText)
End Sub
```

CHAPTER REVIEW

While you were building the word processor, you were introduced to a variety of new Visual Basic features, including many aspects of the Common Dialog control, Visual Basic menus, and the Rich Textbox control. You used a variety of properties and methods of both controls and created menus for the program options.

9 FTP Program

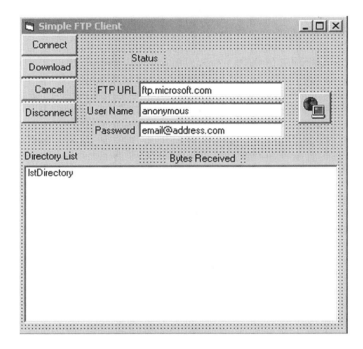

While the intricacies of the File Transfer Protocol (FTP) can be very complicated, developing an FTP client in VB is not. Once you understand the Microsoft Internet Transfer Control (INET) and the functions it provides, the processes involved are actually straightforward and relatively simple.

FTP OVERVIEW

The application we are developing will provide basic access to the FTP process and can be broken up into a series of steps.

1. Connect to FTP server: Try to connect to an FTP server and get a response.
2. Authentication: There are two basic types of connections — anonymous users who access ftp servers for downloading files to their computers and users who have an FTP server in which they have some type of privileges. Either way, you pass a combination of user name and password before the server will allow you to connect.
3. Display File List: Retrieve a list of the files located at the root folder of the FTP server.
4. Change Directory, etc.: Move into the directory of your choice or move onto number 6 if you wish to retrieve a file from the root directory.
5. Retrieve File List: Retrieve a list of files for the directory you changed to.
6. Download File: Download a file from the FTP server to your local system.

Now that you have an idea of the basic process, it's time to begin working on the actual project.

BEGINNING THE PROJECT

Like the controls you have used throughout this book, you need to load the INET control into a project before you can use it. To begin, create a new Standard EXE project in VB. Next, using one of the available methods for accessing the Components window, add the Microsoft Internet Transfer Control to your toolbox and then to the default form.

Basics of the INET Control

The INET Control has useful properties such as RemotePort, URL, Document, UserName and Password, to mention only a few that we are currently interested in. The RemotePort is the port you wish to use to connect to the host, and for FTP, the standard port is 21. Document is the name of a particular file on the FTP server, UserName is your username for the server and Password is the password that corresponds to the username. The last property is URL or the address of the computer you wish to connect with, i.e. ftp://ftp.microsoft.com.

Creating the GUI

Now that we have some of the basics out of the way, we can focus on developing our application, which will connect to an FTP server, retrieve directory and file listings, and download files. For this program we will need to create a series of Command Buttons, Text Boxes, Labels and a List Box. The following list specifies the controls you need and any properties you need to set for each of them:

Control Type	Name	Caption (or Text) Property
Command Button	cmdConnect	Connect
Command Button	cmdDownload	Download
Command Button	cmdCancel	Cancel
Command Button	cmdDisconnect	Disconnect
Label		Bytes Received
Label		Status
Label		FTP URL
Label		User Name
Label		Password
Label		Directory List
Label	lblBytesRec	blank
Label	lblStatus	blank
TextBox	txtURL	ftp.microsoft.com
TextBox	txtUserName	anonymous
TextBox	txtPassword	youremail@yourisp.com
ListBox	lstDirectory	

In the above list, if a control has been given a "blank" entry, you should delete the text information in that field. For example, lblBytesRec should have the text in its caption property deleted. Once you have created the above controls and set their properties, you can arrange them in some manner that resembles Figure 9.1.

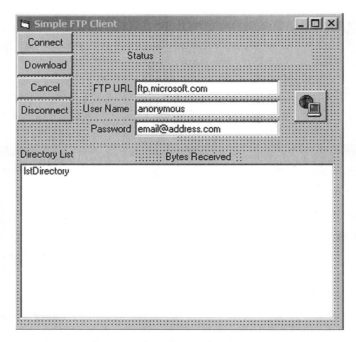

FIGURE 9.1 The completed GUI for the FTP program.

WRITING SOME CODE

Variable Declarations

To begin, let's declare the variables we need for the application. These variables will be explained as they are being used:

```
Option Explicit
Dim inetCMD As String
Dim Data As Variant
Dim strData As String
Dim strTemp As String
Dim lngTemp As Long
Dim cStatus As String
Dim strTemp1 As String
```

Connecting to an FTP Server

The first step required is to connect to the FTP server. The cmdConnect Click event is the location where we will create the code to connect to the server. To begin, we need to set some properties for the Inet1 control beginning with canceling any existing Inet1 operations. Next, we set the following: protocol equal to icFTP, the URL equal to the txtURL textbox, UserName equal to txtUserName and Password equal to txtPassword. Lastly, we need to set the inetCMD variable to list a directory and then execute the command. The following code accomplishes the connection:

```
Private Sub cmdConnect_Click()
 With Inet1
 .Cancel
 .Protocol = icFTP
 .URL = txtURL
 .UserName = txtUserName
 .Password = txtPassword
 End With

 inetCMD = "DIR"
 Inet1.Execute , inetCMD
End Sub
```

There are a few additional points that should be noted with the above code. The use of a With...End With statement simplifies the assignment of the Inet1 properties. Instead of typing Inet1 on every line, we can limit this to a single entry, which reduces time and simplifies the readability of the code. The last line in the procedure executes the command to retrieve the directory. The Execute command will be common throughout the rest of the program when the INET control is being asked to do something. With this in mind, the following is a list of common INET commands:

- CD - Changes directory
- DELETE - Delete a file
- DIR - Retrieve directory listing
- GET - Download a file
- MKDIR - Make directory
- PUT - Uploads file
- PWD - Sends password
- QUIT - Ends session
- RENAME - Renames file
- RMDIR - Deletes directory
- SIZE - Retrieves the size of a file

The vast majority of these commands can only be used if you have advanced rights on an FTP server.

State Changed Event

Once we are connected, we need to have a way to test what is currently happening. Fortunately, the INET control offers a StateChanged event, which is an easy way to monitor what is currently happening with the control. Remember that the inetCMD variable will be a placeholder throughout the program so we can always determine what we are trying to do.

Inside the StateChanged procedure, we will use a series of checks to see what is currently happening. By checking the State value, we can determine the exact position and execution of our program. With this in mind, we'll utilize a Case statement to assign the current status to the lblStatus label on the form. As things change, the label will be updated, offering a convenient way to follow program execution.

The last steps will be to determine if we are doing any special functions like retrieving the directory or a file. If these happen to be the case, we need to take the appropriate actions. The following code is the entire procedure:

```
Private Sub Inet1_StateChanged(ByVal State As Integer)
Select Case State
Case icNone
cStatus = ""
Case icResolvingHost
cStatus = "Resolving Host"
Case icHostResolved
cStatus = "Host Resolved"
Case icConnecting
cStatus = "Connecting"
Case icConnected
cStatus = "Connected"
Case icRequesting
cStatus = "Sending Request"
Case icRequestSent
cStatus = "Request Sent"
Case icReceivingResponse
cStatus = "Receiving Response"
Case icResponseReceived
cStatus = "Response Received"
Case icDisconnecting
cStatus = "Disconnecting"
Case icDisconnected
```

```
    cStatus = "Disconnected"
    Case icError
    cStatus = "Error"
    Case icResponseCompleted
    
    If inetCMD = "DIR" Then
    lstDirectory.Clear
    GetData strData
    lstDirectory.AddItem ".."
    Do While Len(strData)
    lngTemp = InStr(strData, vbCrLf)
    If lngTemp Then
    strTemp = Left$(strData, lngTemp - 1)
    strData = Mid$(strData, lngTemp + 2)
    Else
    strTemp = strData
    strData = ""
    End If
    If Len(Trim$(strTemp)) Then
    lstDirectory.AddItem Trim$(strTemp)
    Loop
    End If
    
    If inetCMD = "GET" Then
    inetCMD = "CDUP"
    Inet1.Execute , inetCMD
    Do While Inet1.StillExecuting: DoEvents: Loop
    inetCMD = "DIR"
    Inet1.Execute , inetCMD
    MsgBox "Download of " & lstDirectory & _
    " is complete.", vbOKOnly
    End If
    
    End Select
    
    lblStatus = cStatus
End Sub
```

You may notice the GetData command being referenced in the Sub procedure. This is a Sub procedure we need to create to use the INET GetChunk method to retrieve data from the FTP server until it equals nothing. We'll use a DoEvents command to temporarily pass control over to Windows and allow it to finish processing events before continuing operations. It's a sort of delay to allow it to catch up. Once is the data from the FTP server is equal to nothing, the procedure will exit. This is the code for the GetData procedure:

```
Private Sub GetData(strData As String)
 Do
 Data = Inet1.GetChunk(256, icString)
 DoEvents
 If Len(Data) = 0 Then Exit Do
 strData = strData & Data
 lblBytesRec.Caption = CStr(Len(strData))
 Loop
End Sub
```

Retrieving Data

Now that we are connected and understand what is taking place, we need to move on to the actual completion of the commands we are giving. The listing of a directory is to occur when we double-click the lstDirectory list box. We can use the event that is raised by this process to enter code for the process.

We begin by setting the strTemp1 variable to the lstDirectory.List(LstDirectory.ListIndex) value, resulting in the variable being assigned to the text information we click on. We'll then pass this information to the inetCMD variable for execution. There is a special circumstance we need to check for. When you double-click on the "..", we need to pass the built-in command "CDUP" rather than the value "..". We can easily accomplish this by beginning the Sub procedure with an If...Then statement to check the value and if it's equal to "..", then execute the "CDUP" command.

Once the command is given, the program needs to loop until completed and once finished, we need to execute the "DIR" command to display the results of the recently changed directory. Here is the code for this procedure:

```
Private Sub lstDirectory_DblClick()
 strTemp1 = lstDirectory.List(lstDirectory.ListIndex)

 If strTemp1 = ".." Then
 inetCMD = "CDUP"
 Inet1.Execute , inetCMD
 End If

 If Right$(strTemp1, 1) = "/" Then
 inetCMD = "CD" & Left$(strTemp1$, Len(strTemp1$) - 1)
 Inet1.Execute , inetCMD
```

```
    End If

    Do While Inet1.StillExecuting: DoEvents: Loop

    inetCMD = "DIR"
    Inet1.Execute , inetCMD

End Sub
```

File Download

Besides retrieving directory listings, we are also hoping to add file retrieval to the program, a task that is even easier than the previous steps. We will use the cmdDownload Click event to change the inetCMD variable to "GET". Next, we simply execute the command with the URL and the filename along with the filename for our computer. For this example, the files will simply be saved in the C root directory with the same name they had on the FTP server. The following code will finish the download process:

```
Private Sub cmdDownload_Click()
  inetCMD = "GET"
  Inet1.Execute txtURL.Text, inetCMD & " " & lstDirectory & _
  " " & "C:\" & lstDirectory
End Sub
```

Finishing Touches

The program is basically finished at this point. However, we need to add a small bit of functionality to cancel the INET command when the cmdCancel button is clicked. Additionally, we need to execute the "QUIT" command when the cmdDisconnect button is clicked and finally, we need to set the INET control to cancel and end the program when the Form_Unload event is raised. Here is the code for those three procedures:

```
Private Sub cmdCancel_Click()
  Inet1.Cancel
End Sub

Private Sub cmdDisconnect_Click()
  Inet1.Execute , "QUIT"
  lstDirectory.Clear
End Sub
```

```
Private Sub Form_Unload(Cancel As Integer)
 Inet1.Cancel
 End
End Sub
```

The program is now finished and can be used for basic FTP functionality. If you'd like to further expand it, you can add additional commands to it. For instance, you know that the "PUT" command functions along the same lines as the "GET" command but instead of retrieving a file, it uploads a file. You could use this to add uploading capabilities to the program in a matter of minutes.

COMPLETE CODE LISTING

The following is the complete code listing for this chapter:

```
Option Explicit

Dim inetCMD As String
Dim Data As Variant
 Dim strData As String
 Dim strTemp As String
 Dim lngTemp As Long
Dim cStatus As String
 Dim strTemp1 As String

Private Sub cmdConnect_Click()
 With Inet1
 .Cancel
 .Protocol = icFTP
 .URL = txtURL
 .UserName = txtUserName
 .Password = txtPassword
 End With

 inetCMD = "DIR"
 Inet1.Execute , inetCMD
End Sub
```

```
Private Sub GetData(strData As String)
 Do
 Data = Inet1.GetChunk(256, icString)
 DoEvents
 If Len(Data) = 0 Then Exit Do
 strData = strData & Data
 lblBytesRec.Caption = CStr(Len(strData))
 Loop
End Sub

Private Sub cmdCancel_Click()
 Inet1.Cancel
End Sub

Private Sub cmdDisconnect_Click()
 Inet1.Execute , "QUIT"
 lstDirectory.Clear
End Sub

Private Sub cmdDownload_Click()
 inetCMD = "GET"
 Inet1.Execute txtURL.Text, inetCMD & " " & lstDirectory & " " & "C:\"
& lstDirectory
End Sub

Private Sub Form_Unload(Cancel As Integer)
 Inet1.Cancel
 End
End Sub

Private Sub Inet1_StateChanged(ByVal State As Integer)
 Select Case State
 Case icNone
 cStatus = ""
 Case icResolvingHost
 cStatus = "Resolving Host"
 Case icHostResolved
 cStatus = "Host Resolved"
 Case icConnecting
 cStatus = "Connecting"
 Case icConnected
 cStatus = "Connected"
 Case icRequesting
 cStatus = "Sending Request"
```

```
    Case icRequestSent
    cStatus = "Request Sent"
    Case icReceivingResponse
    cStatus = "Receiving Response"
    Case icResponseReceived
    cStatus = "Response Received"
    Case icDisconnecting
    cStatus = "Disconnecting"
    Case icDisconnected
    cStatus = "Disconnected"
    Case icError
    cStatus = "Error"
    Case icResponseCompleted

    If inetCMD = "DIR" Then
    lstDirectory.Clear
    GetData strData
    lstDirectory.AddItem ".."
    Do While Len(strData)
    lngTemp = InStr(strData, vbCrLf)
    If lngTemp Then
    strTemp = Left$(strData, lngTemp - 1)
    strData = Mid$(strData, lngTemp + 2)
    Else
    strTemp = strData
    strData = ""
    End If
    If Len(Trim$(strTemp)) Then lstDirectory.AddItem Trim$(strTemp)
    Loop
    End If

    If inetCMD = "GET" Then
    inetCMD = "CDUP"
    Inet1.Execute , inetCMD
    Do While Inet1.StillExecuting: DoEvents: Loop
    inetCMD = "DIR"
    Inet1.Execute , inetCMD
    MsgBox "Download of " & lstDirectory & " is complete.", vbOKOnly
    End If

    End Select

    lblStatus = cStatus
End Sub
```

```
Private Sub lstDirectory_DblClick()
 strTemp1 = lstDirectory.List(lstDirectory.ListIndex)

 If strTemp1 = ".." Then
 inetCMD = "CDUP"
 Inet1.Execute , inetCMD
 End If

 If Right$(strTemp1, 1) = "/" Then
 inetCMD = "CD " & Left$(strTemp1$, Len(strTemp1$) - 1)
 Inet1.Execute , inetCMD
 End If

 Do While Inet1.StillExecuting: DoEvents: Loop

 inetCMD = "DIR"
 Inet1.Execute , inetCMD

End Sub
```

CHAPTER REVIEW

Chapter 9 introduced you to several new aspects of VB programming. First, you learned about the basics of the FTP protocol, and were introduced to the INET control. You learned how to use the With…EndWith statement, which allows you to perform a series of tasks on a specified object without typing the name of the object repeatedly. This helps readability and saves time. Lastly, you learned how the INET control can be given commands to execute some of the FTP functions, such as "DIR" and "GET", to name a few.

10 MDI Web Browser

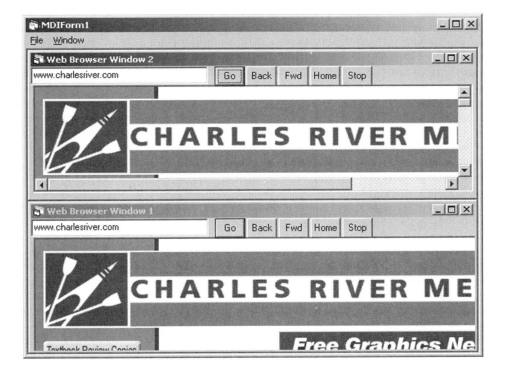

In this chapter we will create a Web Browser based on the Microsoft Internet Control and will introduce you to the concept of a Multiple-Document Interface or MDI, which allows programmers to create applications that can have several documents open at once. Up to this point in the book, all of the applications have been Single Document Interface (SDI); that is, they have been able to open one file at a given time.

MDI OVERVIEW

With an MDI interface, there are any number of child windows and a single parent window (the MDI window). Although child windows can be minimized or maximized, they do so within the constraints of their parent window. To understand the differences between SDI and MDI, you can simply look at Notepad or Microsoft Word, which are displayed in Figures 10.1 and 10.2 respectively. Notepad allows you to open only a single text file at one instance while Word allows you to open several.

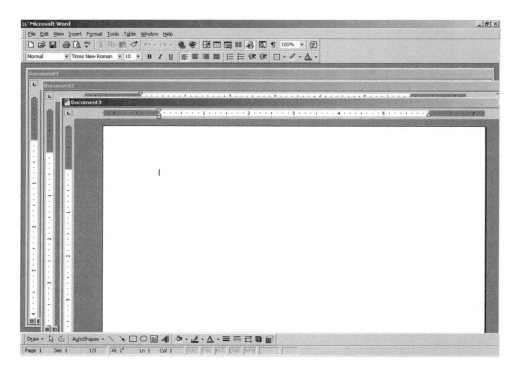

FIGURE 10.1 The Microsoft Word MDI interface allows multiple documents to be open at the same time.

FIGURE 10.2 Notepad only allows access to a single document at any given time.

BEGINNING THE PROJECT

MDI Parent and Child Forms

Creating an MDI project begins much like any VB program. Start Visual Basic and create a Standard EXE program. Next, click on the upper X in the Form window to close the default form that is created with the project but *do not* choose Project| Remove Form1 to remove it, as we'll need it in several steps. We need to add an MDI parent form to our project. Do this by selecting Project|Add MDI Form, then with MDI Form selected in the window that opens, click Open. There should now be two forms shown in the Project Explorer, even though only the MDI form is currently open.

Now that we have an MDI parent, we need to instruct the other form that it needs to be considered an MDI child. This can be accomplished by double-clicking Form1 in the Project Explorer to reopen the default form (or by selecting Form1 from the Window menu), then setting the MDIChild property to True. If you had multiple child forms, you would simply set this property to True on all the child forms. This step changes the child form from SDI to MDI.

The Web Browser we are creating will be capable of displaying several windows at a time. We could create multiple forms by choosing Project|Add Form and set the MDIChild property for all of them to True, but for our application, it will be much easier to create new child forms at runtime when we need them. We'll use Dim form_name As New child_form, which will create new instances of the child form dynamically at runtime.

The MDI child forms that will be created will be slightly different from their SDI counterpart. The following list details some of these differences:

- When a child form is maximized or minimized, it does so within the confines of its parent window rather than the application window.
- When a child form is maximized, its form caption is combined with the form caption of its parent window.
- A child form cannot be hidden – you need to unload it.
- A child form menu will replace the MDI parent menu if it has one; otherwise, the parent menu is displayed.

Creating the Menus

We need to create a simple menu for both the child form and the MDI parent form. We'll begin with the menu structure for the MDI parent:

Create a pull-down menu for the parent form with the following structure:

File
New
Exit

The menu structure on the MDI child is easy as well:

File	*Window*
New	Cascade
Exit	Tile
	Arrange

The Window entry is the only menu item that is out of the ordinary. As you can see in Figure 10.3, the Menu Editor window contains an option called WindowList that can be selected. You should select this option for the Window entry as it automatically creates a list of child windows beneath the Window entry. It displays a complete list of all child windows that are open at a given time, allowing you to easily switch to any of them.

You may have noticed that we created a File menu on the child form as well as on the MDI parent. As I previously mentioned, if the child form has focus, its menu will replace that of the MDI parent.

FIGURE 10.3 The "Window List" option from the Menu Editor automatically creates a list of child windows.

SOME CODE

Dynamic Creation of Child Forms

Let's begin writing code by switching to the parent form and creating the few lines that are needed to create a new child form when the File|New menu item is selected in the MDI parent. The New command will create a new child form and display it inside the MDI parent. The only other code that needs to be placed within the MDI parent is the Exit code for the entire program that will occur when the File|Exit menu item is selected. The following code is all that is needed for both procedures:

```
Private Sub Exit_Click()
 Unload Me
End Sub

Private Sub New_Click()
 Dim frmNew As New frmChild
 frmNew.Show
End Sub
```

At this point in time, if you run the application, and click on the File|New menu item in the parent form, you will create multiple child forms similar to Figure 10.4. You'll also notice that the MDI parent menu is replaced once a child form is in focus. If you close all of the child forms, the default parent Window menu is once again available.

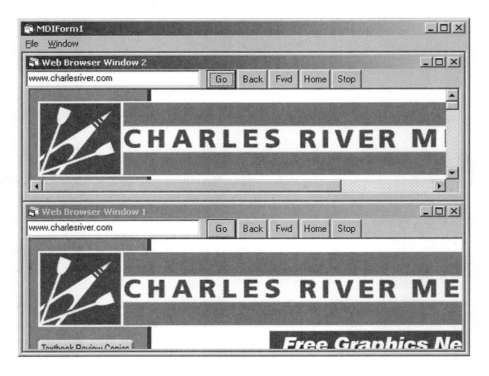

FIGURE 10.4 Multiple child windows can be displayed in the MDI parent.

Child Form Menus

The next step is to switch to the child form. The code that we previously used for File|New and File|Exit on the parent form will continue to work for the child forms, in addition to some new code for arranging windows.

The Window menu with the Cascade, Tile and Arrange commands are very simple to work with. The MDIForm1.Arrange method takes a single integer argument to do each of these tasks. The following list of integers correspond to the listed actions:

Action	Integer
Cascade	0
Horizontal Tile	1
Vertical Tile	2
Arrange Icons	3

Now that you know the values, you can complete the Sub procedures we have been talking about:

```
Private Sub New_Click()
 Dim frmNew As New frmChild
 frmNew.Show
End Sub

Private Sub Close_Click()
 Unload Me
End Sub

Private Sub Cascade_Click()
 MDIForm1.Arrange 0
End Sub

Private Sub Tile_Click()
 MDIForm1.Arrange 1
End Sub

Private Sub ArrangeIcons_Click()
 MDIForm1.Arrange 3
End Sub
```

The Internet Control

At this point in time, we need to have access to the Microsoft Internet Controls. You can add this to your project by choosing Project|Components or by using one of the methods we used earlier. From the list of available components choose Microsoft Internet Controls. The control that is added to your toolbox should look something like a miniature globe and when you position the mouse pointer over the tool, the tool tip will read WebBrowser.

The next step is to double-click this Web Browser control so that it is placed on the form. The small white area will be the area that becomes the Browser Window. You don't need to size it perfectly because we are going to add code that will size it to our exact needs.

The programming commands required to control the Web Browser control are very intuitive. For example, to navigate to a web site, you simply use the Navigate method of the Web Browser control.

The following list gives you some basic ideas of the Web Browser control commands:

- WebBrowser1.Navigate www.yahoo.com: Causes the Yahoo web site to be loaded into the control.
- WebBrowser1.GoBack: Instructs the control to go to the previous address.
- WebBrowser1.Stop: If a site is taking too long, you can stop it.
- WebBrowser1.Refresh: Refreshes the current web site.
- WebBrowser1.GoForward: Causes the browser to revisit a site if you have gone back.
- WebBrowser1.GoHome: Causes your browser to go to its default home page that is set in Internet Explorer.

These are just a few of the basic methods you can utilize to program the control. At this point in time, we'll move on to the programming of our example.

PROGRAMMING THE BROWSER CONTROL

The GUI

A Web Browser requires a few basic pieces of information before it can function, most notably the URL of its destination, and a way to input commands for instructions such as Navigate and Stop. For this example, we'll use a Text Box to enter the destination and a series of command buttons that will enable the user to control the Browser.

The following list of components should be placed on the child form in an arrangement similar to Figure 10.5:

Component	Name	Caption (or Text) Property
TextBox	txtURL	blank
Command Button	cmdGO	Go
Command Button	cmdBack	Back
Command Button	cmdForward	Fwd
Command Button	cmdHome	Home
Command Button	cmdStop	Stop

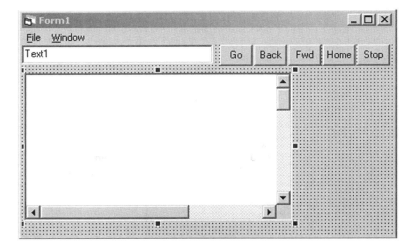

FIGURE 10.5 The Browser window is starting to take shape.

The GUI is complete, so it's time to move forward to the coding of the Web Browser control that will occur as the command buttons are clicked. The events that arise from the button clicks will be the locations of the necessary code.

CREATING THE BROWSER

Web Browser Methods and Properties

I mentioned earlier that the Web Browser control used simple methods to control its functionality. This is noticeable in the following list of completed procedures for the Command Button navigation system we have created:

```
Private Sub cmdGO_Click()
 WebBrowser1.Navigate txtURL.Text
End Sub

Private Sub cmdBack_Click()
 WebBrowser1.GoBack
End Sub

Private Sub cmdForward_Click()
 WebBrowser1.GoForward
End Sub

Private Sub cmdHome_Click()
 WebBrowser1.GoHome
End Sub

Private Sub cmdStop_Click()
 WebBrowser1.Stop
End Sub
```

Resizing the Web Browser Control

If you were to execute the program at this time, it would function as planned but you would notice the limitations that are currently visible. The Web Browser control maintains a very small size, regardless of the size of the window or the web page that is loaded. To handle this, we can simply use the WebBrowser StatusTextChange event to set the Width, Height, and Top position of the Web Browser control. We need to take into account the height of the txtURL text box and subtract this from the values. Lastly, we need to subtract some additional space so as to allow some room for scroll boxes as necessary.

```
Private Sub WebBrowser1_StatusTextChange(ByVal Text As String)

 WebBrowser1.Width = Me.Width - 200
```

```
    WebBrowser1.Height = Me.Height - (txtURL.Top + 350 + 500)
    WebBrowser1.Top = Me.txtURL.Top + 350
End Sub
```

Form Load

The final step for using the Web Browser is to have the caption for the window change dynamically as it is loaded. The following code will accomplish this:

```
Private Sub Form_Load()
  txtURL = "http://www.yahoo.com"
  Me.Caption = "Web Browser Window " & (Forms.Count - 1)
End Sub
```

The final program, which looks similar to Figure 10.6, should now work whenever a child window is resized or a URL is loaded.

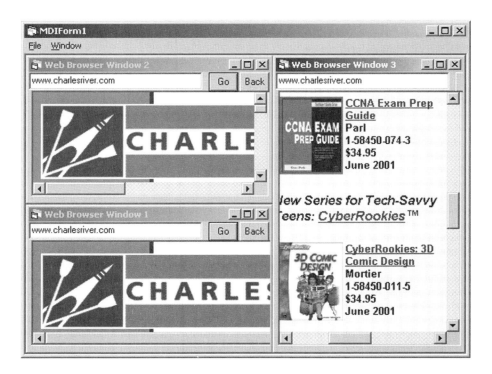

FIGURE 10.6 The final Web Browser with multiple pages open.

COMPLETE CODE LISTING

The following is the complete code listing for this chapter:

Form1:

```
Private Sub ArrangeIcons_Click()
 MDIForm1.Arrange 3 ' Arrange icons
End Sub

Private Sub Cascade_Click()
 MDIForm1.Arrange 0 ' Cascade
End Sub

Private Sub Close_Click()
 Unload Me
End Sub

Private Sub cmdBack_Click()
 WebBrowser1.GoBack
End Sub

Private Sub cmdForward_Click()
 WebBrowser1.GoForward
End Sub

Private Sub cmdGO_Click()
 WebBrowser1.Navigate txtURL.Text
End Sub

Private Sub cmdHome_Click()
 WebBrowser1.GoHome
End Sub

Private Sub cmdStop_Click()
 WebBrowser1.Stop
End Sub

Private Sub Form_Load()
 txtURL = "http://www.yahoo.com"
 Me.Caption = "Web Browser Window " & (Forms.Count - 1)
End Sub
```

```
Private Sub New_Click()
 Dim frmNew As New frmChild form
 frmNew.Show
End Sub

Private Sub Tile_Click()
 MDIForm1.Arrange 1
End Sub

Private Sub WebBrowser1_StatusTextChange(ByVal Text As String)
 WebBrowser1.Width = Me.Width - 200
 WebBrowser1.Height = Me.Height - (txtURL.Top + 350 + 500)
 WebBrowser1.Top = Me.txtURL.Top + 350
End Sub
```

MDIForm1:

```
Private Sub Exit_Click()
 Unload Me
End Sub


Private Sub New_Click()
 Dim frmNew As New frmChild form
 frmNew.Show
End Sub
```

CHAPTER REVIEW

This chapter introduced you to the Microsoft Internet Control, which immediately adds Internet browsing functionality to any form it is placed on by using very simple commands. You learned about several of the commands and how they worked. Lastly, you designed an MDI program and discovered the differences between MDI and SDI.

11 Creating a Chat Program

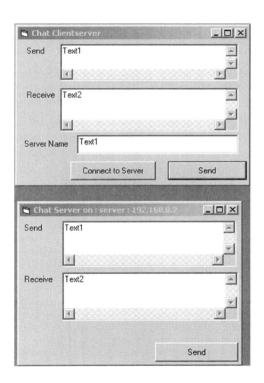

In this chapter we will develop a simple Chat program that uses the Winsock control. But first, in order to program the Winsock control, you need to understand some of the basics of its operation. You'll also need to know the IP address of your local machine to create the program. In order to make sure that any reader of this book can program and test the example, we will develop it in a way to allow it to run both the client and the server on the same PC.

WINSOCK CONTROL

Overview

Like the controls that you are already accustomed to using, it's important to have a basic understanding of the properties and methods that the Winsock control introduces. The following list covers the most important of these properties with an explanation of each:

Properties

- LocalPort: This is the port that your computer will listen to for accessing information as it receives it through the Winsock control. The port can be any four-digit integer number.
- LocalHost: The name of the computer that you are running the application on.
- RemotePort: This is very similar to the local port and is used by the computer you are communicating with to send data. In other words, it's their LocalPort.
- RemoteHost: The name of the computer you are communicating with.
- Protocol: There are two types of protocols that you can choose. The UDP protocol is much slower and tends to lose data packets, but is much easier to program. On the other hand, TCP is much faster and more reliable, but is much more difficult to program.
- LocalIP: Your IP address.
- SocketHandle: Returns the handle of the Socket, which can be passed, to API calls.
- State: Returns the state of the connection.
- BytesReceived: The total number of bytes you have received from the moment you connected to another computer.

Methods

- Accept: Accepts a request for connection.
- Bind: Binds the control to a specific port and adapter and is used if you have multiple protocol adapters.
- Close: Closes the current connection.
- Connect: Attempts to connect to the specified PC.
- GetData: Gets the data that has arrived and clears the data buffer.
- Listen: Listens on the specified port.

- PeekData: Similar to GetData; looks at the information without removing it from the buffer.
- SendData: Sends data once a connection has been established.

CREATING THE FORMS

Server Form

The first step in creating the chat program is to create a new Standard EXE file in VB. Next, add the WinSock control to your toolbar by choosing Project|Components and selecting the check box next to Microsoft Winsock Control. Change the name of the default form that was created with the project to frmServer in the Properties window. The form's caption will be set dynamically at runtime.

The following controls should be added to frmServer:

Control Type	Name	Caption or Text	Additional
Text Box	txtSend	blank	MultiLine=True, ScrollBars=3-Both
Text Box	txtReceive	blank	MultiLine=True, ScrollBars=3-Both
CommandButton	cmdSend	Send	
Winsock	tcpServer		

You can add labels next to the Text Boxes so that you understand which is Send and which is Receive. The form should look similar to Figure 11.1. You may notice that the Text Boxes have their MultiLine property set equal to True, and their ScrollBars property set equal to both Horizontal and Vertical. This allows large messages to be sent and a line feed to be used between messages.

The Client Form

The client form is almost identical to the server form. Add a new form to the project by choosing Project|AddForm and rename it to frmClient. The form's caption will be set dynamically at runtime. Add the following controls to it:

Control Type	Name	Caption or Text	Additional
Text Box	txtSend	blank	MultiLine=True, ScrollBars=3-Both
Text Box	txtReceive	blank	MultiLine=True, ScrollBars=3-Both
TextBox	txtPCName	blank	
Command Button	cmdSend	Send	
Command Button	cmdConnect	Connect to Server	
Winsock	tcpClient		

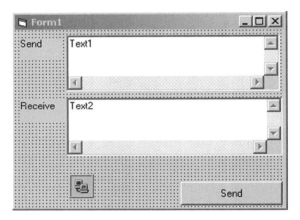

FIGURE 11.1 The server form for our program.

Like the server form, you can add labels to identify the Text Boxes if you'd like, so that it looks like Figure 11.2.

Startup Form

The last form we need to create is a form that will display the client and server forms and set their positions on the screen when a command button is clicked. Add a form to the project with the name frmStart and add a Command Button, changing its name to cmdOpenForms and its Caption to Open Client and Server. Next, choose Project|Properties and select frmStart as the Startup form. The form should look something like Figure 11.3.

You can use the Show method for both the client and server forms to display them, and once on the screen, you can position them so they are not overlapping. This startup form should then unload itself from memory, as the only function it offers is to provide a way to display the other forms.

Here is the code for frmStart:

```
Private Sub cmdOpenForms_Click()
 frmClient.Show
 frmClient.Left = 100
 frmClient.Top = 100

 frmServer.Show
 frmServer.Left = 100
 frmServer.Top = 5000
 Unload Me
End Sub
```

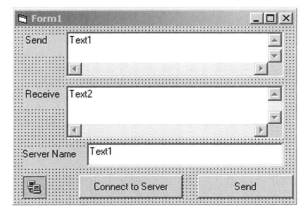

FIGURE 11.2 The layout of the client form is very similar to that of the server form.

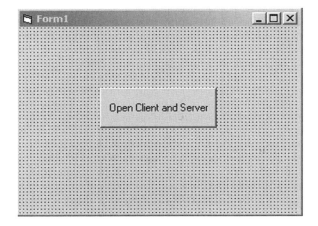

FIGURE 11.3 The startup form is very simple and is only used to display the client and server forms.

WRITING CODE FOR THE CLIENT AND SERVER

Server

Now that basic forms have been created, it's time to write some code for the server. To begin, we need to switch to the server form and set the LocalPort property of the Winsock control to 100, which is the port that the server will listen to. I've chosen a number that is easy to remember, but it can be changed to another integer as long as the client knows what this value is. Next, we need to use the Listen property to instruct the server to listen on that port. We also need to set the caption for the server form.

The following code can be added to the Form_Load procedure of frmServer:

```
Private Sub Form_Load()
 tcpServer.LocalPort = 100
 tcpServer.Listen
 Me.Caption = "Chat Server on : " & tcpServer.LocalHostName & _
 " : " & tcpServer.LocalIP
End Sub
```

We also need to inform the Winsock control what it needs to do when a connection is requested. There are two simple steps that need to occur. The first determines if there is a current connection, and if so, it should be stopped in favor of the new request. Second, we need to accept the new request.

```
Private Sub tcpServer_ConnectionRequest(ByVal requestID As Long)
 If tcpServer.State <> sckClosed Then tcpServer.Close
 tcpServer.Accept requestID
End Sub
```

A few additional procedures are needed to handle the Send button being clicked and the handling of the data that will arrive from the client. The following procedures handle these functions using the GetData and SendData methods and the use of the strData variable and txtSend and txtReceive Text Boxes:

```
Private Sub tcpServer_DataArrival(ByVal bytesTotal As Long)
 Dim strData As String
 tcpServer.GetData strData, vbString
 txtReceive.Text = txtReceive.Text & vbCrLf & strData
End Sub

Private Sub cmdSend_Click()
```

```
      tcpServer.SendData txtSend.Text
End Sub
```

 In data arrival, vbcrlf is a built-in vb command used for a carriage return-line feed.

Client

The server is now complete and we can focus our attention on the client, a much easier prospect, as the code is similar with only a simple variable name change from tcpServer to tcpClient. Here is the code for frmClient:

```
Private Sub Form_Load()
 tcpClient.RemotePort = 100
 Me.Caption = "Chat Client" & tcpClient.LocalHostName
End Sub

Private Sub tcpClient_DataArrival(ByVal bytesTotal As Long)
 Dim strData As String
 tcpClient.GetData strData, vbString
 txtReceive.Text = txtReceive.Text & vbCrLf & strData
End Sub

Private Sub cmdSend_Click()
 tcpClient.SendData (txtSend.Text)
End Sub

Private Sub cmdConnect_Click()
 tcpClient.RemoteHost = txtPCName
 tcpClient.Connect
End Sub
```

TESTING THE PROGRAM

To test the program, you should run it by clicking Start on the toolbar within the VB IDE. When the frmStart form is displayed, you should click the Open Client and Server button. This will display both forms in an arrangement similar to Figure 11.4. Next, add the Hostname or IP address of the server to the client form's Server Name Text Box (txtPCName) and click the Connect to Server button. Once you have connected, feel free to type a message from the client in the Send Text Box and click the Send button and if everything is OK, you should immediately receive it on the server in the Receive Text Box. You can try this in the other direction as well.

If the application runs correctly on a single PC and you have access to a network (or via the Internet), you can test it by running the application and only opening the client on one PC and the server on the other. You could also choose to remove frmStart and create two separate programs.

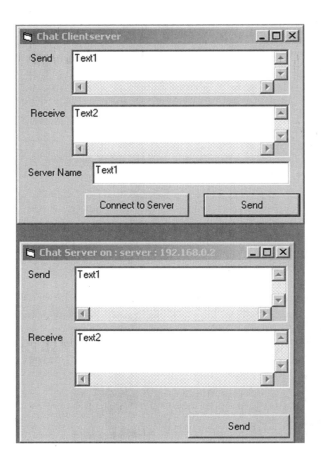

FIGURE 11.4 The client and server forms are displayed.

COMPLETE CODE LISTING

The following code is a complete listing for this chapter:

frmClient:

```
Option Explicit

Private Sub Form_Load()
 tcpClient.RemotePort = 100
 Me.Caption = "Chat Client" & tcpClient.LocalHostName
End Sub

Private Sub tcpClient_DataArrival(ByVal bytesTotal As Long)
 Dim strData As String
 tcpClient.GetData strData, vbString
 txtReceive.Text = txtReceive.Text & vbCrLf & strData
End Sub

Private Sub cmdSend_Click()
 tcpClient.SendData (txtSend.Text)
End Sub

Private Sub cmdConnect_Click()
 tcpClient.RemoteHost = txtPCName
 tcpClient.Connect
End Sub
```

frmServer:

```
Option Explicit

Private Sub Form_Load()
 tcpServer.LocalPort = 100
 tcpServer.Listen
 Me.Caption = "Chat Server on : " & tcpServer.LocalHostName & _
 " : " & tcpServer.LocalIP
End Sub

Private Sub tcpServer_ConnectionRequest(ByVal requestID As Long)
 If tcpServer.State <> sckClosed Then tcpServer.Close
 tcpServer.Accept requestID
End Sub
```

```
Private Sub tcpServer_DataArrival(ByVal bytesTotal As Long)
 Dim strData As String
 tcpServer.GetData strData, vbString
 txtReceive.Text = txtReceive.Text & vbCrLf & strData
End Sub

Private Sub cmdSend_Click()
 tcpServer.SendData txtSend.Text
End Sub
```

frmStart:

```
Option Explicit

Private Sub cmdOpenForms_Click()
 frmClient.Show
 frmClient.Left = 100
 frmClient.Top = 100

 frmServer.Show
 frmServer.Left = 100
 frmServer.Top = 5000

 Unload Me
End Sub
```

CHAPTER REVIEW

In this chapter, you were introduced to the Winsock control, and some of the properties and methods it offers. Once you had some basic information, we created a client and server chat program that is capable of operating on the same PC, or can be used on PCs across a network. Lastly, you were provided with some basic information on how to deal with multiple forms in a single program.

12 Animated Desktop Assistant

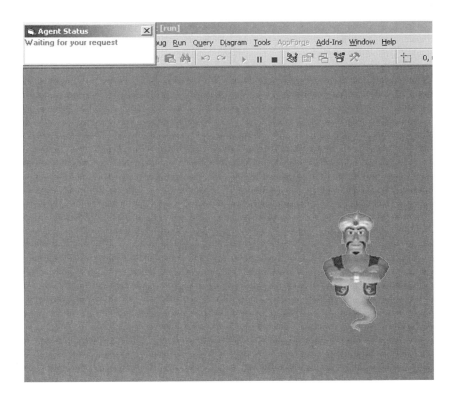

This chapter introduces you to the Microsoft Agent control. Once you are comfortable with the basics of the control, we will develop an animated Desktop Assistant which can be used to open applications, visit your favorite web site, or read you the current date and time utilizing Text-To-Speech and Voice Recognition Commands.

AGENT OVERVIEW

Microsoft Agent is a freely available control that supports the presentation of interactive animated characters within Windows and includes support for speech recognition, which allows your programs to respond to voice commands, along with mouse and keyboard input. The characters in turn respond, using synthesized speech or optional text in a cartoon style word balloon. One of the standard Microsoft characters named Genie can be seen in Figure 12.1.

FIGURE 12.1 Agent characters can convert text information to speech output.

Installing Microsoft Agent

ON THE CD

For your convenience, all of the necessary Microsoft Agent tools have been placed on the CD-ROM included with this book. These packages include the Agent chore components, the Text-To-Speech (TTS) engine developed by Lernout & Hauspie, Microsoft SAPI (Speech API), Microsoft Speech Recognition Engine, and the Speech Control Panel. They also include the four standard Microsoft characters, which can be seen in Figure 12.2, including Peedy, Merlin, Genie and Robby. If you are interested, Microsoft provides tools on the Agent web site for designing your own Agents or you can download many freely available Agents on the Internet.

The tools are located in the X:\Programs\Agent directory where X is the letter of your CD-ROM drive. If Agent is currently not on your system, you should install all of the Agent files before you begin the programming exercises contained in this chapter. Additional languages and tools are available at the Microsoft Agent web site, which is located at

http://msdn.microsoft.com/workshop/c-frame.htm#/workshop/imedia/Agent/default.asp.

FIGURE 12.2 The standard Microsoft Agent characters.

CREATING THE DESKTOP ASSISTANT

To begin this project, create a new Standard EXE application within Visual Basic. Next, you need to determine which of the alternatives you will use to gain access to Microsoft Agent. The first, and easiest solution, is to place a Microsoft Agent ActiveX control on a form. The other alternative is to declare an Agent control reference at runtime. The second option, although more difficult, does offer a single advantage. If you develop an application with Agent, and distribute it, the second option will allow your application to run correctly, albeit without Agent support. If you use the first method and run the same application on another computer, it will display error messages and may crash or simply stop executing. In this exercise, however, we will use the first method.

Placing the Control on Your Form

In order to place the Agent control on a Visual Basic form, you will first need to add the Agent component to your VB toolbox. Using the shortcut combination of Ctrl+T, right-clicking the toolbox, or using the Project|Components command, select the Microsoft Agent Control from the list of available controls. Once it is in your toolbox, the Agent icon will look something like a cartoon secret agent. Double-click the icon to the form and place it anywhere you would like. It is not displayed at runtime, so the exact position is unimportant.

Writing Some Code

Next, double-click on the form, which will bring up the Code window. The Code window should be empty with the exception of the Form_Load subprocedure. We need to add several variable declarations, which will be used as placeholders for the center of the screen, and for Agent. Next, we'll load the Agent Character file, which defines what your Agent will sound and look like. It contains the default TTS information and all of the character animations.

Because the Agent information is contained in the character file, you need to specify what character file to open before you begin programming with Agent. We'll use the Agent1.Characters.Load command with the Genie character being specified. It is then displayed with the Show method. You should normally check for the existence of a character, but because we are using the Genie character that is included on the CD-ROM, you don't have a need for this type of error checking.

```
Private Genie As IAgentCtlCharacterEx
Dim CenterX As Integer
Dim CenterY As Integer

Private Sub Form_Load()
 Agent1.Characters.Load "Genie", "Genie.acs"
 Set Genie = Agent1.Characters("Genie")
 Genie.Show
End Sub
```

DECLARING AN AGENT CONTROL REFERENCE AT RUNTIME

The rest of this chapter will assume that you have placed the Agent control on your form using the methods that have been mentioned up to this point in time. However, you can create a reference to it at runtime instead which provides an advantage in error handling related to individuals running your application without having Agent installed. This information is being included so that you can decide which method you prefer to use in your own custom applications. Do not type any of the code for this section, but use it for future reference.

In order to take advantage of the increased error handling provided by this option, an On Error Resume Next statement would be used, which would prevent the program from being stopped by an error caused by a computer that does not have the Agent control installed. In our example, we would not actually check the error to make sure what was causing the problem. In fact, the error would be the same if the user didn't have the Genie character installed or if they didn't have Agent installed at all.

We would need to create a few variables that will store the default Agent installation locations (c:\windows\msagent\chars\) so that we could attempt to load Genie. We would also need a variable for the Agent character.

The next step would be to create the On Error Resume Next statement. Next, the ctrlAgent variable should then be set to an Agent control instance with Set ctrlAgent = CreateObject("Agent.Control.1"). The ctrlAgent.Connected property would need to be set equal to True. The rest of the code would be identical to the current example.

```
Option Explicit

Dim strLocalPath As String
Dim strAgentName As String
Dim ctrlAgent

Private Sub Form_Load()
 On Error Resume Next
 strLocalPath = "c:\windows\msagent\chars\"
 strAgentName = "genie"
 Set ctrlAgent = CreateObject("Agent.Control.1")
 ctrlAgent.Connected = True
 ctrlAgent.Characters.Load strAgentName, strLocalPath & strAgentName & ".acs"
 ctrlAgent.Characters(strAgentName).Show
End Sub
```

ADDING SPEECH AND ANIMATION

Animation

If you run the application at this point, your program will open the Genie Agent and display it on the screen in a manner similar to Figure 12.3. Now that you have the character appearing on the screen, you need to get to do something with it. With Microsoft Agent, this can be accomplished very quickly and easily.

FIGURE 12.3 The Genie Agent appearing on your desktop.

By using methods that are available with the Agent characters, we can instruct the Agent to play animation files such as "wave". At the Agent web site, Microsoft supplies a list of all standard animation files and an animation tool that allows you to query an individual Agent to not only see the names and number of animations, but also to view them. It is very important to know the exact name of an animation, as the application will function erratically if the Agent is given a command to play an animation that is invalid.

To program the Text-To-Speech aspect of Agent, we can use the Speak method. This is a simple one-line command; all you need to provide is a string. It will convert it to speech and display it in the word balloon without any additional programming required. Like the majority of TTS engines, the speech output tends to be given without emotion or reflection, but you can modify the way words are spoken to give it more of a real life sound using speech output tags.

Speech output tags can be used to modify the way words are spoken and can accomplish such tasks as emphasizing certain words or making the Agent whisper. The tags start and end with a backslash, and modify the word following the output tag. In our example, we'll add the speech tag \emp\ tag immediately before the Agent is instructed to say its own name. When you execute the sample, you should notice the pitch of the Agent's voice as it changes.

Enter the following code after the Genie.Show command that has already been entered:

```
Genie.Play "Wave"
Genie.Speak "Hello, my \emp\name is " & Genie.Name & "."
Genie.Speak "I am at your service."
```

The agent characters have two basic types of animations: looping and non-looping. Generally speaking, looping animations will continue functioning until you stop them. Stopping a looping animation is very easy, and can be accomplished by using the stop command (ie. Genie.stop).

Speech Recognition

Your Desktop Agent program can display the Genie Agent and convert text information to speech, but at this point in time it is merely a novelty, as it does not perform any really useful tasks. The application will use the Speech Recognition Engine to give the Agent a command to execute when a certain word is spoken.

Microsoft Agent provides the necessary tools to respond to speech input, but you must first add a command or a series of commands that can be carried out by the Agent. By requiring commands to be given in this manner, the Agent's accuracy dramatically increases, as the Speech Recognition Engine does not have to compare your spoken command to every word in the spoken language. These commands are simply a word, or a series of words. The Agent does not constantly listen for commands, but instead provides a keyboard key that activates the Agent. By default, the Agent key utilizes the SCROL LLOCK key.

You could add the commands in the Form_Load subprocedure, but for our example, you will create a custom subprocedure. This subprocedure, which will be given the name AddCommands, will be used by our application to separate the commands from the rest of the program. To begin, type the following after the End Sub of the Form Load () subprocedure.

```
Private Sub AddCommands()

End Sub
```

Adding Commands

The Microsoft Agent parameters for adding commands are actually quite simple and contain several options. In order to comprehend the command, you must first become acquainted with the built-in Agent subprocedure called Agent_Command. This subprocedure is called every time a command is given to the Agent either by right-clicking and selecting the command from the Voice Commands pop-up window which can be seen in Figure 12.4, or by voice input. It is in this subprocedure that we will later place our code for determining which command is being given to the Agent.

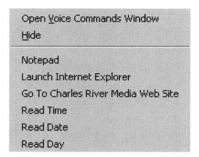

FIGURE 12.4 The Agent Voice Commands window.

The basic syntax of a typical Agent command is as follows:

```
Genie.Commands.Add "Notepad", "Notepad", " Notepad", True, True
```

In the above example, the Genie.Commands.Add method does exactly what its name implies – it adds a command to the Genie Agent. The "Notepad" portion of the line, which is known as the Name parameter, will be returned in the Agent1_Command() event. The second parameter, "Open Notepad", is the Caption property parameter that is displayed in the Agent Voice Commands window which is displayed when you right-click on the Agent. The third parameter, "Execute Notepad", is the Voice property parameter, which is what you must say in order to execute this command. As you can see in the example, the Voice property does not have to be the same as the Name or the Caption properties.

You can add the following commands to your AddCommands subprocedure above the words End Sub:

```
With Genie.Commands
  .Add "Notepad", "Notepad", "Notepad", True, True
  .Add "Explorer", "Launch Internet Explorer", "Launch Explorer", True, True
  .Add "Charles River Media", "Go To Charles River Media Web Site", "Charles River Media", True, True
  .Add "Time", "Read Time", "Time", True, True
  .Add "Date", "Read Date", "Date", True, True
  .Add "Today", "Read Day", "Today", True, True
End With
```

If you execute the application, and right-click on the Genie, you may notice a couple of commands that you did not create. Microsoft Agent comes with a few built-in commands like "Open" and "Hide" which respectively show or hide the Agent. These commands do not need to be added to the Agent Voice Commands window.

The commands we have created will be used to do some fairly obvious tasks such as the first command: "Notepad" will be used to open Notepad. The second command will be used to open Internet Explorer, the third will be used to open Internet Explorer and visit the Charles River Media Web Site, and the fourth, fifth, and sixth commands will be used to read the time, date and day respectively. The commands will not actually perform any tasks at this time, but you'll add some code later in the chapter to handle the tasks associated with them.

Enter the following code after the End With statement :

```
Form1.Left = 0
Form1.Top = 0
 With Screen
 CenterX = Int((.Width / .TwipsPerPixelX) / 2)
 CenterY = Int((.Height / .TwipsPerPixelY) / 2)
 End With
Genie.MoveTo CenterX, CenterY
```

This code begins by moving the form to the top left side of the screen. It also determines the center of the screen using the TwipsPerPixelX and TwipsPerPixelY Visual Basic commands. Finally, we move the Genie to the center of the screen with the MoveTo method.

THE GUI

Next, we need to create a Label control on the form so that it appears similar to Figure 12.5. This Label control will be used to display some basic information about what our Agent is currently doing. You can also change the caption property of the form to Agent Status.

FIGURE 12.5 The simple GUI for our application.

Add the following line, which is used to set the label's default value to "Waiting for your request", to the end of the Form_Load() Event :

```
Label1 = "Waiting for your request"
```

You have finished the AddCommands subprocedure, but you need to call it from within the program. The following call can be placed after the SetGenie = Agent1.Characters ("Genie") command in the Form_Load event:

```
AddCommands
```

EXECUTING THE COMMANDS

As we mentioned previously, a command can be given to the Agent by right-clicking and selecting it from the pop-up menu or by the Microsoft Speech Recognition Engine. The latter requires the use of a quality microphone, and even then, the Agent has to determine which of the commands you were actually after. Because speech input is a very difficult process, this is not always completely accurate, so the commands should be used only when they can be verified, or for items that are not extremely important. For instance, it is probably not a good idea to have the Agent delete files from your hard drive.

When an Agent command is given, it calls the Agent_Command subprocedure with UserInput passed as an object of Agent_Command. This object allows us to test how much Confidence the Agent has in its selection. We can test this Confidence value, which ranges from −100 to 100, making sure that we are getting something that is accurate. The Code window provides an easy way to create the Agent1_Command subprocedure. Select Agent1 from the Object drop-down menu followed by Command from the Procedure drop-down menu.

In the Agent1_Command subprocedure, type the following code:

```
If UserInput.Confidence < -25 Then
  Genie.Speak "I'm sorry, I cannot determine the command!"
  Label1.Caption = "Cannot determine the command"
  Exit Sub
End If
```

As you can see by the code, the UserInput Confidence is compared with −25. If it is less than −25, the Agent will respond with "I'm sorry, I cannot determine the command." It also sets the label on your form to reflect that it cannot determine the command, and exits the procedure so that Agent does not simply guess at your request. The –25 value is something that can be adjusted for your particular needs.

If you choose to add additional recognition features to an agent application, there are parameters that can be used to let you know what other commands the agent considered when it attempted to determine your voice input. You can get the names of the other commands agent was considering by using the userinput.alt1name property and the userinput.alt2name property. For example, these parameters could be used to prompt a user to choose which of the commands they actually wanted.

You are nearing completion of your Desktop Assistant, but at this time, the Agent does not know what to do with the commands it receives. Add the following code after the End If statement you previously created:

```
Select Case UserInput.Name
 Case "Notepad"
 Shell "Notepad.exe", vbNormalFocus
 Case "Explorer"
 Shell "C:\Program Files\Internet Explorer\iexplore.exe",_
vbNormalFocus
 Case "Charles River Media"
 Shell "C:\Program Files\Internet Explorer\iexplore.exe _
http://www.charlesriver.com", vbNormalFocus
 Case "Time"
 If Hour(Now) > 12 Then
 Genie.Speak "The time is " & Hour(Now) - 12 & ":" & _
Minute(Now) & " PM"
 Else
 Genie.Speak "The time is " & Hour(Now) & ":" & Minute(Now) & _
" AM"
 End If
 Case "Today"
 Genie.Speak "Today is " & WeekdayName(Weekday(Now))
 Case "Date"
 Genie.Speak "Today is " & WeekdayName(Weekday(Now)) & _
", " & MonthName(Month(Now)) & " " & Day(Now) & ", " _
 & Year(Now)
 Case Else
 Genie.Speak "The command, " & UserInput.Voice & ", is unknown!"
End Select
Label1.Caption = "The command " & UserInput.Name & " was executed."
```

Perhaps the best way to check a series of possibilities is by using the Case statement. The above segment utilizes the Case statement to check the value of UserInput.Name, thereby determining the command that was given. For example, the first line checks to see if the command was "Notepad". If it is, it uses the Shell command to execute Notepad.exe. If the command is not "Notepad" it continues by checking the next option, which in the example, is the "Explorer" command. The other possible commands include the Day, Date and Time, which utilize the built-in VB commands for day, date, and time functions. If it continues through all possibilities without a match, the Agent informs the user that they didn't understand the command. Lastly, we set the Label1.Caption to UserInputName.

If you run the application, you should have a completed Desktop Assistant that looks similar to Figure 12.6 and is capable of opening applications, visiting web sites, or telling the time. It could be completely customized to provide a tremendous amount of time-saving features. For instance, you could hide the form as its only purpose is to provide user feedback. This would leave the Agent alone on your desktop. You could then show and hide the Agent as needed, creating an Assistant that was visible only when you wanted a task performed.

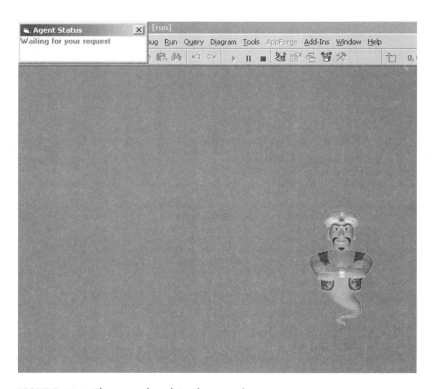

FIGURE 12.6 The completed Desktop Assistant.

COMPLETE CODE LISTING

The following is a complete code listing for this chapter:

```
Option Explicit

Private Genie As IAgentCtlCharacterEx
Dim CenterX As Integer
Dim CenterY As Integer

Private Sub Form_Load()
 Agent1.Characters.Load "Genie", "Genie.acs"
 Set Genie = Agent1.Characters("Genie")
 AddCommands
 Genie.Show
 Genie.Play "Wave"
 Genie.Speak "Hello, my \emp\name is " & Genie.Name & "."
 Genie.Speak "I am at your service."
 Label1 = "Waiting for your request"
End Sub

Private Sub AddCommands()
 With Genie.Commands
 .Add "Notepad", "Notepad", "Notepad", True, True
 .Add "Explorer", "Launch Internet Explorer", "Launch Explorer", True, True
 .Add "Charles River Media", "Go To Charles River Media Web Site", "Charles River Media", True, True
 .Add "Time", "Read Time", "Time", True, True
 .Add "Date", "Read Date", "Date", True, True
 .Add "Today", "Read Day", "Today", True, True
 End With

 Form1.Left = 0
 Form1.Top = 0
 With Screen
 CenterX = Int((.Width / .TwipsPerPixelX) / 2)
 CenterY = Int((.Height / .TwipsPerPixelY) / 2)
 End With
 Genie.MoveTo CenterX, CenterY
End Sub

Private Sub Agent1_Command(ByVal UserInput As Object)
If UserInput.Confidence < -25 Then
```

```
      Genie.Speak "I'm sorry, I cannot determine the command!"
      Label1.Caption = "Cannot determine the command"
      Exit Sub
    End If
    Select Case UserInput.Name
      Case "Notepad"
        Shell "Notepad.exe", vbNormalFocus
      Case "Explorer"
        Shell "C:\Program Files\Internet Explorer\iexplore.exe", vbNormalFocus

      Case "Charles River Media"
        Shell "C:\Program Files\Internet Explorer\iexplore.exe
    http://www.charlesriver.com", vbNormalFocus
      Case "Time"
        If Hour(Now) > 12 Then
          Genie.Speak "The time is " & Hour(Now) - 12 & ":" & Minute(Now) & " PM"
        Else
          Genie.Speak "The time is " & Hour(Now) & ":" & Minute(Now) & " AM"
        End If
      Case "Today"
        Genie.Speak "Today is " & WeekdayName(Weekday(Now))
      Case "Date"
        Genie.Speak "Today is " & WeekdayName(Weekday(Now)) & ", " &
    MonthName(Month(Now)) & " " & Day(Now) & ", " & Year(Now)
      Case Else
        Genie.Speak "The command, " & UserInput.Voice & ", is unknown!"
    End Select

    Label1.Caption = "The command " & UserInput.Name & " was executed."
End Sub
```

CHAPTER REVIEW

In this chapter, you were introduced to several new features, most notably the Microsoft Agent control. Along with the basics of the Microsoft Agent control, you learned about the methods it provides and how to add voice input and commands to your Agent. We also looked at the Case statement and how beneficial it is to use it to check a series of values. Lastly, we looked at how we can execute a program outside of Visual Basic using the Shell command.

13 Creating an Internet Time Retrieving ActiveX Control

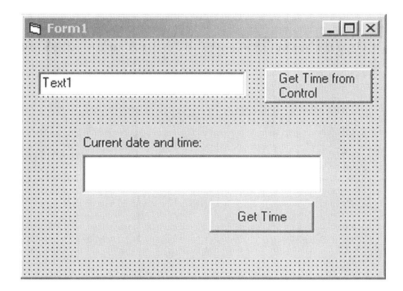

In the previous chapters of this book we have used a wide range of ActiveX controls in the projects we've created. You can see the obvious advantage of using the controls as they provide a set of functions that you would otherwise be required to code yourself. In this chapter, we'll develop our own custom ActiveX control that will offer properties like most ActiveX controls. The basic function of the ActiveX control will be to retrieve the current date and time from Internet web sites using the Winsock control.

ACTIVEX PROJECT

The GUI

Like most things in Visual Basic, creating ActiveX controls is a straightforward process and not overly difficult. To start the project, open Visual Basic but instead of choosing Standard EXE, you need to choose the "Create an ActiveX Control" option. Your screen, which should appear similar to Figure 13.1, looks something like a form without a border. If you think about it, this makes complete sense as the vast majority of Visual Basic controls do not have a border.

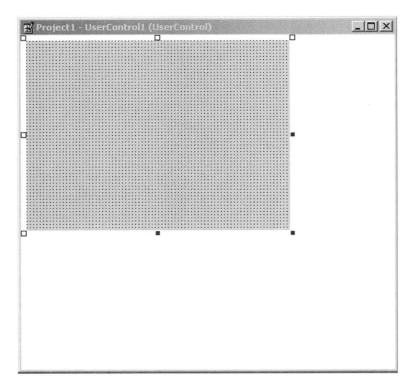

FIGURE 13.1 ActiveX controls are developed much like any VB project.

We need to add the following controls to the User Control, which is named UserControl1, and arrange them in a manner similar to Figure 13.2. (Remember to get the Winsock control from the Components window.)

Creating an Internet Time Retrieving ActiveX Control

FIGURE 13.2 The ActiveX control interface.

Control Type	Name	Caption
Winsock	Winsock1	
CommandButton	cmdGetTime	Get Time
TextBox	txtDateTime	blank
Label		Current date and time:

Developing the Control

As you have probably noticed, developing the GUI for an ActiveX control is nearly identical to developing the GUI for a regular VB EXE application. Writing the code is nearly identical as well. There are a few minor differences, but we'll discuss them as we encounter them.

When a regular Visual Basic form loads, a Form_Load event is raised. On the other hand, ActiveX controls offer a UserControl_Initialize event that works the same. For this application, double-click on the user control to display the Code window. Using the UserControl_Initialize event, you need to set txtDateTime equal to blank space. Here is the Initialize code:

```
Private Sub UserControl_Initialize()
 txtDateTime = ""
End Sub
```

Next, we need to add some code to the cmdGetTime_Click event. We begin by setting txtDateTime equal to blank space (if the user executes this code more than once, it's important to erase the previous date and time). We should close any previous Winsock connections and set the local port equal to 0. Lastly, we connect to an Internet server, in our case pogostick.net, which offers the ability to retrieve the date and time.

The following code will complete the procedure:

```
Private Sub cmdGetTime_Click()
 txtDateTime = ""
 Winsock1.Close
 Winsock1.LocalPort = 0
 Winsock1.Connect "pogostick.net", 13
End Sub
```

You'll notice that we are using Port 13 in the last line of the procedure. Port 13 allows us simple access to the Network Time Protocol or NTP. NTP was devised as a method of retrieving the date and time from other computers and was originally developed to allow networked computers to synchronize dates and times amongst themselves. The technology continues to be popular. If you would prefer, you can do an Internet search for NTP servers and substitute one of them for pogostick.net.

As you may remember, the Winsock control provides a DataArrival event that allows us to capture the data that it is receiving. We'll use this event in a manner very similar to the Chat example in Chapter 11. To begin, we create a variable for storing the received data and then assign a Text Box to this variable. For this example, we'll use the Trim$ function to trim any leading or trailing spaces, leaving us with only the data. Here is the code for this procedure:

```
Private Sub Winsock1_DataArrival(ByVal bytesTotal As Long)
 Dim strData
 Winsock1.GetData strData, vbString
 txtDateTime = txtDateTime & Trim$(strData)
End Sub
```

We can add some simple error handling to this as well. The Winsock1_Error event will return an error code and description if an error occurs. We'll simply use a Message Box to inform the end user of any problem. Here is the code:

```
Private Sub Winsock1_Error(ByVal Number As Integer, Description As String,_
 ByVal Scode As Long, ByVal Source As String, ByVal HelpFile As String, _
 ByVal HelpContext As Long, CancelDisplay As Boolean)
 MsgBox "ActiveX Control Winsock Error: " & Number & vbCrLf & Description, vbInformation
End Sub
```

Properties

Like other controls, we want the ActiveX control to provide a set of properties so that the end user can modify values or receive information from the control. For our particular control, we need to have access to the time that will be placed in txtDateTime.

Creating properties for ActiveX controls is easy but at first can seem a little confusing. For instance, when you set a property equal to something, you are running one piece of code. When you retrieve a property, you are running a totally different piece of code. This sometimes confuses beginning ActiveX programmers, but actually makes sense if you stop to consider it.

To add a new property to an ActiveX control, you need to follow these steps:

- Open the Code window for the control.
- Select Tools|Add Procedure.
- In the Name box, type your property name (for our example, "ITime").
- Click the Property option button.
- Click OK.

You should notice the following added to your Code window:

```
Public Property Get ITime() As Variant

End Property

Public Property Let ITime(ByVal vNewValue As Variant)

End Property
```

The two properties correspond to either assigning a value (Let) or retrieving a value (Get). For our control, we can delete the Let property, as we only need to retrieve information from the control. It wouldn't really seem to make sense to assign a value from your program to a control that is supposed to retrieve a value from the Internet. We should also change the Get property to type String, as we'll be assigning it to a Text Box when we test it.

Once that is finished, we simply assign the value of txtDateTime equal to ITime. Here is the code for the property:

```
Public Property Get ITime() As String
  ITime = txtDateTime.Text
End Property
```

TESTING THE CONTROL

Creating Another Project

To test the ActiveX control, we cannot simply run it in the IDE like we do most Visual Basic applications. Instead, we have to create a project that will utilize the control. It is the additional project that we can then execute to test the functions of our ITime property.

We could compile the ActiveX control and then place it inside a project but rather than go through all of the steps, we can test its functions inside VB. First, we need to add a new project to the current ActiveX project. To do this, close the Code and Form windows for the ActiveX control. Next, select File|Add New Project. You can select Standard EXE in the New Project Window.

On the default form, you can place a Text Box and Command Button with the following properties:

Control Type	Name	Caption or Text
TextBox	txtGetTime	Text1
Command Button	cmdGetFromCtl	Get Time from Control

Next, you can choose our UserControl1 from the toolbox and place it on the form as well, changing its name to InternetTime. Your form should look similar to Figure 13.3.

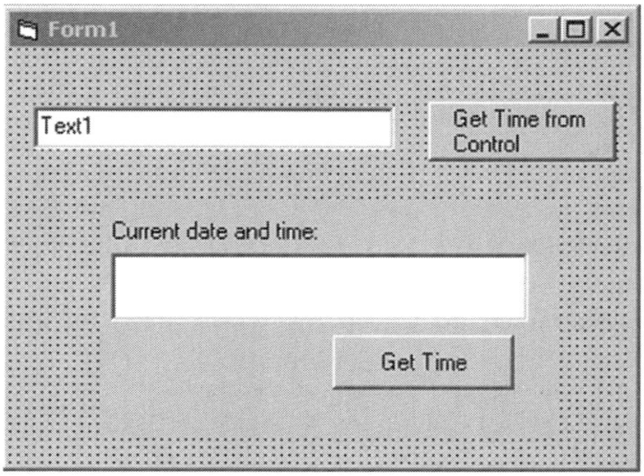

FIGURE 13.3 When it is placed on the form, your ActiveX control should look like any standard control.

We need to add some code that will retrieve the ITime property from the InternetTime control. If we did everything correctly in the previous steps, you should be able to add the following code to the cmdGetFromControl Click Event:

```
Private Sub cmdGetFromCtl_Click()
 txtGetTime = InternetTime.Itime
End Sub
```

That's the only line necessary for this to function. If necessary, connect to your Internet Service Provider, then switch to your project. You can now execute the project like any regular VB program. Click the Get Time button inside the control we created. Within a few seconds, you should see a date and time visible inside the text box located in the control. Now, when you click the Get Time From Control button, you should see the date and time visible in the txtGetTime TextBox.

COMPLETE CODE LISTING

The following is the complete code listing for this chapter:

UserControl:

```
Private Sub cmdGetTime_Click()
 txtDateTime = ""
 Winsock1.Close
 Winsock1.LocalPort = 0
 Winsock1.Connect "pogostick.net", 13
End Sub

Private Sub UserControl_Initialize()
 txtDateTime = ""
End Sub

Private Sub Winsock1_DataArrival(ByVal bytesTotal As Long)

 Dim strData

 Winsock1.GetData strData, vbString
 txtDateTime = txtDateTime & Trim$(strData)
End Sub
```

```
Private Sub Winsock1_Error(ByVal Number As Integer, Description As
String, ByVal Scode As Long, ByVal Source As String, ByVal HelpFile As
String, ByVal HelpContext As Long, CancelDisplay As Boolean)

 MsgBox "ActiveX Control Winsock Error : " & Number & vbCrLf &
Description, vbInformation
End Sub

Public Property Get Itime() As String
 Itime = txtDateTime.Text
End Property
```

Standard EXE Project:

```
Private Sub cmdGetFromCtl_Click()
 txtGetTime = InternetTime.ITime
End Sub
```

CHAPTER REVIEW

This chapter covered a wide range of topics, mostly related to the creation of ActiveX controls within VB. You learned how to retrieve and set properties for your ActiveX control using Get and Let. We combined the information we learned about the Winsock control in a previous chapter and used it to retrieve the date and time from the Internet using NTP on Port 13.

14 Visual Basic Paint Program

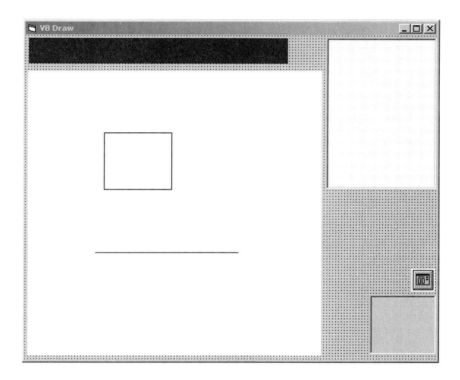

In this chapter, we will develop a Paint program using a wide range of topics we have already covered, such as the API and the intrinsic controls like the Picture Box. We'll look at alternative ways for designing toolbars, and we will create a pop-up menu using a variation on a standard menu. Lastly, we'll learn how to create an Undo feature for the application that could be altered to work with most VB programs.

THE PAINT PROJECT

GUI

To begin this project, we need to create a standard EXE file in VB. On the default form we need to place several controls, in an arrangement similar to Figure 14.1, with the following properties:

Control Type	Name	BackColor
PictureBox	picDraw	
PictureBox	picUndo	
PictureBox	picToolbar	&H00FF0000& (Blue—Palette)
PictureBox	picToolbar2	&H80000018& (ToolTip—System)
CommonDialog1	CommonDialog1	
Shape	Rect1	
Line	Line1	

You also need to adjust the controls according to the following list:

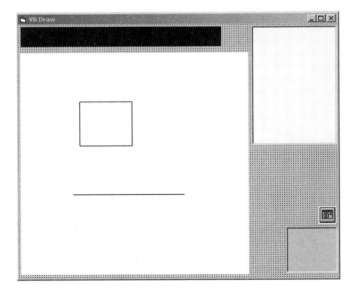

FIGURE 14.1 The beginnings of the user interface.

Control	Left	Width	Height
picDraw	60	6015	6015
picUndo	7080	1335	1215
picToolbar	75	5295	540
picToolbar2	6180	2235	3210

Toolbars

ON THE CD

You may have noticed that the two Picture Box Controls with altered background colors are named picToolbar and picToolbar2. We'll use them, along with the API, to make toolbars that can be moved at runtime by the end user. The first toolbar (picToolbar) will contain a series of Command Buttons that will be used for the common drawing functions like Line, Circle, Rectangle, Fill and Undo to name a few. Instead of using text information to display the button functions, we'll use icons, which can be set in the Properties Window using the Picture property. You can use the icons, which appear in Figure 14.2 and are included on the CD-ROM in the Figures directory. Alternately, you can design your own. Add the following Command Buttons to picToolbar and change their Names as shown. Change the Picture property for each Command Button to the appropriate icon from the CD-ROM, using Figure 14.2 as a reference.

Name		
cmdPaint	cmdEllipse	cmdErase
cmdLine	cmdFill	cmdMove
cmdRectangle	cmdDropper	cmdUndo
cmdFilledRectangle	cmdText	

FIGURE 14.2 The first toolbar contains most of the commands for the program.

The next step is the second toolbar, picToolbar2. It contains the following controls:

Control Type	Name	Fill Color
PictureBox	picColor	
Shape	shpLineColor	&H00000000& (Black)
ComboBox	cmbDrawStyle	
ComboBox	cmbWidth	
ComboBox	cmbCanvasWidth	
ComboBox	cmbCanvasHeight	

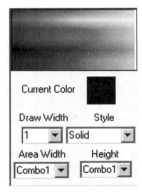

FIGURE 14.3 The second of the toolbars should be arranged similar to this figure.

ON THE CD

Along with the controls listed above, you need to place labels for the Shape and the Combo Boxes and change their captions to match Figure 14.3. You'll probably notice that the picColor Picture Box control has had a picture loaded into it from the CD-ROM. The figure can be found in the Figures\Chapter 14 directory and is named colors.jpg. Lastly, you need to change cmbWidth's Text property to 1.

Pop-Up Menu

The pop-up menu begins by creating a standard VB menu. Create the menu with the follow structure and later we'll cover the code that turns this ordinary menu into a pop-up menu:

Menu Caption	Name
PopUp	mnuPopUp
...&File	Menu
......&New	mnuNew
......&Open	mnuOpen
......&Save	mnuSave
...&Exit	mnuExit

WRITING THE CODE

API and Variable Declarations

We'll begin writing code for this application by setting up the API declarations and variables. The API declarations are listed below with some basic information about each of them:

API	Information
ReleaseCapture	Ends any mouse capture that is in effect.
SendMessage	Sends a message to a window.
Ellipse	Draws an ellipse.

This is the exact syntax for the API declarations and can be entered into the General Declarations section of the Code window:

```
Option Explicit
```

```
Private Declare Function ReleaseCapture Lib "user32" () As Long

Private Declare Function SendMessage Lib "user32" Alias _
"SendMessageA" (ByVal hWnd As Long, ByVal wMsg As Long, _
ByVal wParam As Long, lParam As Any) As Long

Private Declare Function Ellipse Lib "Gdi32" (ByVal hdc As Long, _
  ByVal X1 As Long, ByVal Y1 As Long, ByVal X2 As Long, _
  ByVal Y2 As Long) As Long
```

There are eight variables and two constants that need to be created as well. They keep track of mouse button clicks and the current tool being used. They also track the positions of the mouse on the picDraw PictureBox for starting, old, and current X and Y positions.

You can enter this also into the General Declarations section of the Code window:

```
Const WM_NCLBUTTONDOWN = &HA1
Const HTCAPTION = 2
Dim blnChangeColor As Boolean
Dim CurrentTool As String
Dim CurrX As Integer
Dim CurrY As Integer
Dim OldX As Integer
Dim OldY As Integer
Dim StartX As Integer
Dim StartY As Integer
```

FORM LOAD

We begin writing code for this application in the Form_Load event. The application will provide the ability to change drawing styles with a click on the cmbDrawStyle Combo Box. With this in mind, we need to add drawing styles to the cmbDrawStyle Combo Box by using its AddItem method.

Next, we use a For…Next loop to add a series of numbers to the cmbWidth (changes the width of the tool you are using), cmbCanvasWidth, and cmbCanvasHeight Combo Box. We will set the Visible property of mnuPopUp to False so that is not seen on the form at the top of the screen, and we set the current Width and Height properties of the cmbCanvasWidth and cmbCanvasHeight ComboBoxes equal to the current width and height of the picDraw PictureBox.

Here is the code for this procedure:

```
Private Sub Form_Load()
Dim I As Integer
 cmbDrawStyle.AddItem "Solid"
 cmbDrawStyle.AddItem "Dash"
 cmbDrawStyle.AddItem "Dot"
 cmbDrawStyle.AddItem "DashDot"
 cmbDrawStyle.AddItem "DashDotDot"
 cmbDrawStyle.AddItem "Transparent"
 cmbDrawStyle.AddItem "Transp. Solid"
 cmbDrawStyle.ListIndex = "0"
 picDraw.DrawStyle = cmbDrawStyle.ListIndex

 For I = 1 To 50
 cmbWidth.AddItem I
 Next I

 For I = 1 To 480
 cmbCanvasWidth.AddItem I
 Next I

 For I = 1 To 640
 cmbCanvasHeight.AddItem I
 Next I

 mnuPopUp.Visible = False

 cmbCanvasHeight = Int(picDraw.Height / 15)
 cmbCanvasWidth = Int(picDraw.Width / 15)
End Sub
```

MENU EVENTS

Once the form has been loaded and the variables have been initialized, we can focus our attention on the menu events. There are only four commands: New, Open, Save and Exit. The Open and Save options use the CommonDialog control and filters much like the VB word processor did in an earlier chapter. The Exit command will exit the program using Unload, and the New command will clear the picDraw and picUndo Picture Boxes. Here is the code for the procedures:

```
Private Sub mnuExit_Click()
 Unload Me
End Sub

Private Sub mnuNew_Click()
 picDraw.Cls
 picUndo.Cls
End Sub

Private Sub mnuOpen_Click()
 CommonDialog1.CancelError = True
 On Error GoTo ErrHandler
 CommonDialog1.Filter = "Windows bitmap (*.BMP)|*.bmp| _
GIF" & "(*.gif)|*.gif|JPEG Filter|*.jpg|All|*.*"
 CommonDialog1.FilterIndex = 1
 CommonDialog1.ShowOpen
 picDraw.Picture = LoadPicture(CommonDialog1.FileName)
 Exit Sub
ErrHandler:
 Exit Sub
End Sub

Private Sub mnuSave_Click()
 CommonDialog1.CancelError = True
 On Error GoTo ErrHandler
 CommonDialog1.Filter = "Windows bitmap (*.BMP)|*.bmp| _
GIF" & "(*.gif)|*.gif|JPEG Filter|*.jpg|All|*.*"
 CommonDialog1.FilterIndex = 1
 CommonDialog1.ShowSave
 SavePicture picDraw.Image, CommonDialog1.FileName
 Exit Sub
ErrHandler:
 Exit Sub
End Sub
```

CHANGING COLORS

Changing colors is a normal requirement for a painting program. We can use the picColor Picture Box, which has a color selector picture already loaded, and then use the MouseDown, MouseMove and MouseUp events to determine the color that is being selected. We can then pass this information to the shpLineColor.FillColor, picDraw.ForeColor, and set the Boolean value blnChangeColor equal to True.

The reason we need to use MouseDown, MouseMove and MouseUp events is to allow the end user of the program to simply click the picture or click and drag across the color selector and see their color changes instantaneously. The code for these three procedures can be entered now:

```
Private Sub picColor_MouseDown(Button As Integer, Shift As Integer, X As Single, Y As Single)
On Error Resume Next
 shpLineColor.FillColor = picColor.Point(X, Y)
 picDraw.ForeColor = picColor.Point(X, Y)
 blnChangeColor = True
End Sub

Private Sub picColor_MouseMove(Button As Integer, Shift As Integer, X As Single, Y As Single)
On Error Resume Next
 If blnChangeColor Then
 shpLineColor.FillColor = picColor.Point(X, Y)
 picDraw.ForeColor = picColor.Point(X, Y)
 End If
End Sub

Private Sub picColor_MouseUp(Button As Integer, Shift As Integer, X As Single, Y As Single)
 blnChangeColor = False
End Sub
```

WIDTH, HEIGHT, AND STYLE

The project has a user-selected width and height for picDraw, the space that the user will actually draw on. These changes are made when the cmbCanvasWidth and cmbCanvasHeight boxes are changed. We also have a width available for the current drawing tool that, for example, allows the width of a line to be user-selectable. Lastly, we need to set the Style of the current drawing tool from the cmbDrawStyle ComboBox. The following code can be entered now:

```
Private Sub cmbCanvasWidth_Click()
 picDraw.Width = Int(cmbCanvasWidth.ListIndex * 15)
End Sub

Private Sub cmbCanvasHeight_Click()
 picDraw.Height = Int(cmbCanvasHeight.ListIndex * 15)
End Sub

Private Sub cmbWidth_Click()
 picDraw.DrawWidth = cmbWidth.ListIndex
 Rect1.BorderWidth = cmbWidth.ListIndex
 Line1.BorderWidth = cmbWidth.ListIndex
End Sub

Private Sub cmbDrawStyle_Click()
 picDraw.DrawStyle = cmbDrawStyle.ListIndex
End Sub
```

Moving the Toolbars

Rather than using static toolbars, we have used Picture Boxes to act as a frame for our menu items. Using Picture Boxes will give us the option to move the toolbars around the screen at runtime. We will use the MouseDown events on both toolbars and the ReleaseCapture and SendMessage APIs to complete the moves. Here is the necessary code:

```
Private Sub picToolbar2_MouseDown(Button As Integer, _
Shift As Integer, X As Single, Y As Single)
 ReleaseCapture
 SendMessage picToolbar2.hWnd, WM_NCLBUTTONDOWN, HTCAPTION, 0&
End Sub

Private Sub picToolbar_MouseDown(Button As Integer, Shift As Integer, X
As Single, Y As Single)
 ReleaseCapture
 SendMessage picToolbar.hWnd, WM_NCLBUTTONDOWN, HTCAPTION, 0&
End Sub
```

THE COMMANDS

The next step in our application is to use the Command Buttons in picToolbar to assign the current drawing tool. We will simply set the CurrentTool variable equal to a text representation of the tool. This value will later be used in the drawing procedure.

Here is the code for all of the Command Button Click events:

```
Private Sub cmdDropper_Click()
 CurrentTool = "Dropper"
End Sub

Private Sub cmdEllipse_Click()
 CurrentTool = "Ellipse"
End Sub

Private Sub cmdErase_Click()
 CurrentTool = "Erase"
End Sub

Private Sub cmdFilledRectangle_Click()
 CurrentTool = "FilledRectangle"
End Sub

Private Sub cmdLine_Click()
 CurrentTool = "Line"
End Sub

Private Sub cmdMove_Click()
 CurrentTool = "Move"
End Sub

Private Sub cmdFill_Click()
 CurrentTool = "Fill"
End Sub

Private Sub cmdPaint_Click()
 CurrentTool = "Paint"
End Sub

Private Sub cmdRectangle_Click()
 CurrentTool = "Rectangle"
End Sub
```

```
Private Sub cmdText_Click()
 CurrentTool = "Text"
End Sub
```

Undo

The Undo procedure is the final toolbar option that needs to be addressed. The Undo command will simply set the picDraw Picture property equal to the picUndo Picture property. The picUndo PictureBox is updated a step behind the picDraw Picture Box, allowing you to go back one entry. You could create additional Picture Box controls to add greater levels of Undo functionality. Here is the code for the procedure:

```
Private Sub cmdUndo_Click()
 picDraw.Picture = picUndo.Picture
End Sub
```

DRAWING TO THE PICTUREBOX

The project is nearly finished as you only have three procedures left to complete. They are the picDraw MouseDown, MouseMove and MouseUp procedures, which are collectively responsible for drawing on the PictureBox.

MouseDown Event

The MouseDown event is the first of the three that we'll look at. It will begin by setting the picUndo picture property equal to picDraw. As I previously mentioned, this will allow us to have a single remove a single element from the drawing at any given time. Next, we update OldX, OldY, StartX, and StartY. These values are very important for drawing to the exact points on the Picture Box. We then use a series of If…Then statements to check the CurrentTool for specific string values.

Depending on the value, we need to perform certain functions. For instance, if the value is equal to "Fill" then we will set the picDraw.DrawStyle property to vbSolid and then draw a rectangle over the entire Picture Box. If the value is equal to "Move", we can use the SendMessage and ReleaseCapture APIs to move the Picture Box like we previously did with the toolbars. A "Dropper" value indicates that the Dropper tool has been selected and we need to change colors using the picDraw Picture Box like we did earlier with picColor. There is different code being executed for each of the tools.

The final check for this procedure is to see if the button clicked is the right mouse button. A button value is passed to the MouseDown procedure as an integer, with the right button having a value of 2. If the button being clicked is the right button, we need to display your pop-up menu by using the PopUpMenu command.

The following procedure can be added to your Code window:

```
Private Sub picDraw_MouseDown(Button As Integer, Shift As Integer, X As Single, Y As Single)
 picUndo.Picture = picDraw.Image ' Set for Undo before Any Drawing

 OldX = X
 OldY = Y
 StartX = X
 StartY = Y
 If CurrentTool = "Fill" Then
 picDraw.LineStyle = vbSolid
 picDraw.Line (0, 0)-(pictDraw.Width, pictDraw.Height), , BF
 picDraw.Refresh
 End If
 If CurrentTool = "Move" Then
 ReleaseCapture
 SendMessage picDraw.hWnd, WM_NCLBUTTONDOWN, HTCAPTION, 0&
 End If
 If CurrentTool = "Dropper" Then
 blnChangeColor = True
 shpLineColor.FillColor = picDraw.Point(X, Y)
 picDraw.ForeColor = picDraw.Point(X, Y)
 End If
 If Button = 2 Then
 PopupMenu mnuPopUp
 End If
End Sub
```

MouseMove and MouseUp Events

The MouseMove and MouseUp procedures work nearly identical to the MouseDown procedure. They perform a series of checks on the value of CurrentTool. Depending on this value, a certain type of line, rectangle, ellipse or text is drawn to picDraw. There is very little to mention about these procedures, as most of the items have been already covered in this chapter.

One of the few new items is the actual drawing of shapes. By using the OldX, OldY and the current X and Y values, almost any type of imaginable shape can be created. For most of these shapes, you can use a built-in picDraw method, such as line. That said, the one item that cannot be drawn this way is the ellipse which uses a call to the Windows API (Ellipse picDraw.hdc, OldX, OldY, X, Y).

The rest of the procedures should be straightforward enough to figure out:

```
Private Sub picDraw_MouseMove(Button As Integer, Shift As Integer, X As Single, Y As Single)

Select Case CurrentTool
Case "Paint":
If Button = 1 Then
picDraw.Line (OldX, OldY)-(X, Y), shpLineColor.FillColor
OldX = X
OldY = Y
End If
Case "Line":
If Button = 1 Then
Line1.X1 = OldX
Line1.Y1 = OldY
Line1.X2 = X
Line1.Y2 = Y
Line1.Visible = True
End If
Case "Rectangle":
If Button = 1 Then
Rect1.Visible = True
Rect1.BorderColor = picDraw.ForeColor
Rect1.FillStyle = 1 ' Transparent
If X > OldX Then
Rect1.Width = X - OldX
Rect1.Left = OldX
End If
If Y > OldY Then
Rect1.Height = Y - OldY
Rect1.Top = OldY
End If
If X < OldX Then
Rect1.Left = X
Rect1.Width = OldX - X
End If
If Y < OldY Then
Rect1.Top = Y
```

```
Rect1.Height = OldY - Y
End If
End If
Case "Ellipse":
If Button = 1 Then
End If
Case "Dropper"
If blnChangeColor = True Then
shpLineColor.FillColor = picDraw.Point(X, Y)
picDraw.ForeColor = picDraw.Point(X, Y)
End If
Case "FilledRectangle":
If Button = 1 Then
Rect1.Visible = True
Rect1.FillColor = picDraw.ForeColor
Rect1.BorderColor = picDraw.ForeColor
Rect1.FillStyle = 0 'Solid
If X > OldX Then
Rect1.Width = X - OldX
Rect1.Left = OldX
End If
If Y > OldY Then
Rect1.Height = Y - OldY
Rect1.Top = OldY
End If
If X < OldX Then
Rect1.Left = X
Rect1.Width = OldX - X
End If
If Y < OldY Then
Rect1.Top = Y
Rect1.Height = OldY - Y
End If
End If
Case "Erase":
If Button = 1 Then
picDraw.Line (OldX, OldY)-(X, Y), picDraw.BackColor
OldX = X
OldY = Y
End If

End Select
End Sub
```

```vb
Private Sub picDraw_MouseUp(Button As Integer, Shift As Integer, X As Single, Y As Single)
 Debug.Print CurrentTool, OldX, X
 Select Case CurrentTool
 Case "Line":
 picDraw.Line (OldX, OldY)-(X, Y)
 Line1.Visible = False
 Case "Ellipse":
 Call Ellipse(picDraw.hdc, OldX, OldY, X, Y)
 picDraw.Refresh
 Case "Rectangle":
 picDraw.Line (OldX, OldY)-(X, Y), , B
 Rect1.Visible = False
 Case "FilledRectangle":
 picDraw.Line (OldX, OldY)-(X, Y), , BF
 Rect1.Visible = False
 Case "Dropper"
 shpLineColor.FillColor = picDraw.Point(X, Y)
 picDraw.ForeColor = picDraw.Point(X, Y)
 blnChangeColor = False
 Case "Text"
 CommonDialog1.Flags = cdlCFScreenFonts
 CommonDialog1.ShowFont
 On Error GoTo ErrorHandler 'If user cancels Common Dialog
 picDraw.FontName = CommonDialog1.FontName
 picDraw.FontSize = CommonDialog1.FontSize
 picDraw.FontBold = CommonDialog1.FontBold
 picDraw.FontItalic = CommonDialog1.FontItalic
 picDraw.FontUnderline = CommonDialog1.FontUnderline
 picDraw.FontStrikethru = CommonDialog1.FontStrikethru
 picDraw.CurrentX = X
 picDraw.CurrentY = Y
 picDraw.Print InputBox("Please Enter Your Text", "Text Tool")
ErrorHandler:
 End Select
End Sub
```

COMPLETE CODE LISTING

The following code is the complete listing for this chapter:

```
Option Explicit

Private Declare Function ReleaseCapture Lib "user32" () As Long
Private Declare Function SendMessage Lib "user32" Alias "SendMessageA"
(ByVal hWnd As Long, ByVal wMsg As Long, ByVal wParam As Long, lParam
As Any) As Long
Private Declare Function Ellipse Lib "Gdi32" (ByVal hdc As Long, ByVal
X1 As Long, ByVal Y1 As Long, ByVal X2 As Long, ByVal Y2 As Long) As
Long

Const WM_NCLBUTTONDOWN = &HA1
Const HTCAPTION = 2

Dim blnChangeColor As Boolean
Dim CurrentTool As String
Dim CurrX As Integer
Dim CurrY As Integer
Dim OldX As Integer
Dim OldY As Integer
Dim StartX As Integer
Dim StartY As Integer

Private Sub cmbCanvasWidth_Click()
 picDraw.Width = Int(cmbCanvasWidth.ListIndex * 15)
End Sub

Private Sub cmbCanvasHeight_Click()
 picDraw.Height = Int(cmbCanvasHeight.ListIndex * 15)
End Sub

Private Sub cmbDrawStyle_Click()
 picDraw.DrawStyle = cmbDrawStyle.ListIndex
End Sub

Private Sub cmbWidth_Click()
 picDraw.DrawWidth = cmbWidth.ListIndex
 Rect1.BorderWidth = cmbWidth.ListIndex
 Line1.BorderWidth = cmbWidth.ListIndex
End Sub
```

```vb
Private Sub cmdDropper_Click()
 CurrentTool = "Dropper"
End Sub

Private Sub cmdEllipse_Click()
 CurrentTool = "Ellipse"
End Sub

Private Sub cmdErase_Click()
 CurrentTool = "Erase"
End Sub

Private Sub cmdFilledRectangle_Click()
 CurrentTool = "FilledRectangle"
End Sub

Private Sub cmdLine_Click()
 CurrentTool = "Line"
End Sub

Private Sub cmdMove_Click()
 CurrentTool = "Move"
End Sub

Private Sub cmdFill_Click()
 CurrentTool = "Fill"
End Sub

Private Sub cmdPaint_Click()
 CurrentTool = "Paint"
End Sub

Private Sub cmdRectangle_Click()
 CurrentTool = "Rectangle"
End Sub

Private Sub cmdText_Click()
 CurrentTool = "Text"
End Sub

Private Sub cmdUndo_Click()
 picDraw.Picture = picUndo.Picture
End Sub
```

```
Private Sub Form_Load()

Dim I As Integer
 cmbDrawStyle.AddItem "Solid"
 cmbDrawStyle.AddItem "Dash"
 cmbDrawStyle.AddItem "Dot"
 cmbDrawStyle.AddItem "DashDot"
 cmbDrawStyle.AddItem "DashDotDot"
 cmbDrawStyle.AddItem "Transparent"
 cmbDrawStyle.AddItem "Transp. Solid"
 cmbDrawStyle.ListIndex = "0"
 picDraw.DrawStyle = cmbDrawStyle.ListIndex

 For I = 1 To 50
 cmbWidth.AddItem I
 Next I

 For I = 1 To 480
 cmbCanvasWidth.AddItem I
 Next I

 For I = 1 To 640
 cmbCanvasHeight.AddItem I
 Next I

 mnuPopUp.Visible = False

 cmbCanvasHeight = Int(picDraw.Height / 15)
 cmbCanvasWidth = Int(picDraw.Width / 15)
End Sub

Private Sub mnuExit_Click()
 Unload Me
End Sub

Private Sub mnuNew_Click()
 picDraw.Cls
 picUndo.Cls
End Sub

Private Sub mnuOpen_Click()

 CommonDialog1.CancelError = True
 On Error GoTo ErrHandler
```

```
  CommonDialog1.Filter = "Windows bitmap (*.BMP)|*.bmp| _
GIF" & "(*.gif)|*.gif|JPEG Filter|*.jpg|All|*.*"


  CommonDialog1.FilterIndex = 1

  CommonDialog1.ShowOpen


  picDraw.Picture = LoadPicture(CommonDialog1.FileName)

  Exit Sub

ErrHandler:
 'User pressed the Cancel button
 Exit Sub

End Sub

Private Sub mnuSave_Click()

  CommonDialog1.CancelError = True
  On Error GoTo ErrHandler



  CommonDialog1.Filter = "Windows bitmap (*.BMP)|*.bmp| _
GIF" & "(*.gif)|*.gif|JPEG Filter|*.jpg|All|*.*"



  CommonDialog1.FilterIndex = 1

  CommonDialog1.ShowSave


  SavePicture picDraw.Image, CommonDialog1.FileName
  Exit Sub

ErrHandler:

  Exit Sub
```

```
End Sub

Private Sub picDraw_MouseDown(Button As Integer, Shift As Integer, _
 X As Single, Y As Single)
 picUndo.Picture = picDraw.Image
 
 OldX = X
 OldY = Y
 StartX = X
 StartY = Y
 If CurrentTool = "Fill" Then
 picDraw.LineStyle = vbSolid
 picDraw.Line (0, 0)-(picDraw.Width, picDraw.Height), , BF
 picDraw.Refresh
 End If
 If CurrentTool = "Move" Then
 ReleaseCapture
 SendMessage picDraw.hWnd, WM_NCLBUTTONDOWN, HTCAPTION, 0&
 End If
 If CurrentTool = "Dropper" Then
 blnChangeColor = True
 shpLineColor.FillColor = picDraw.Point(X, Y)
 picDraw.ForeColor = picDraw.Point(X, Y)
 End If
 If Button = 2 Then
 PopupMenu mnuPopUP
 End If

End Sub

Private Sub picDraw_MouseMove(Button As Integer, Shift As Integer, X As
Single, Y As Single)

 Select Case CurrentTool
 Case "Paint":
 If Button = 1 Then
 picDraw.Line (OldX, OldY)-(X, Y), shpLineColor.FillColor
 OldX = X
 OldY = Y
 End If
 Case "Line":
 If Button = 1 Then
 Line1.X1 = OldX
 Line1.Y1 = OldY
```

```
Line1.X2 = X
Line1.Y2 = Y
Line1.Visible = True
End If
Case "Rectangle":
If Button = 1 Then
Rect1.Visible = True
Rect1.BorderColor = picDraw.ForeColor
Rect1.FillStyle = 1 ' Transparent
If X > OldX Then
Rect1.Width = X - OldX
Rect1.Left = OldX
End If
If Y > OldY Then
Rect1.Height = Y - OldY
Rect1.Top = OldY
End If
If X < OldX Then
Rect1.Left = X
Rect1.Width = OldX - X
End If
If Y < OldY Then
Rect1.Top = Y
Rect1.Height = OldY - Y
End If
End If
Case "Ellipse":
If Button = 1 Then
End If
Case "Dropper"
If blnChangeColor = True Then
shpLineColor.FillColor = picDraw.Point(X, Y)
picDraw.ForeColor = picDraw.Point(X, Y)
End If
Case "FilledRectangle":
If Button = 1 Then
Rect1.Visible = True
Rect1.FillColor = picDraw.ForeColor
Rect1.BorderColor = picDraw.ForeColor
Rect1.FillStyle = 0 'Solid
If X > OldX Then
Rect1.Width = X - OldX
Rect1.Left = OldX
End If
```

```
        If Y > OldY Then
        Rect1.Height = Y - OldY
        Rect1.Top = OldY
        End If
        If X < OldX Then
        Rect1.Left = X
        Rect1.Width = OldX - X
        End If
        If Y < OldY Then
        Rect1.Top = Y
        Rect1.Height = OldY - Y
        End If
        End If
        Case "Erase":
        If Button = 1 Then
        picDraw.Line (OldX, OldY)-(X, Y), picDraw.BackColor
        OldX = X
        OldY = Y
        End If
        
        End Select
    End Sub
    
    Private Sub picDraw_MouseUp(Button As Integer, Shift As Integer, _
     X As Single, Y As Single)
     Debug.Print CurrentTool, OldX, X
     Select Case CurrentTool
     Case "Line":
     picDraw.Line (OldX, OldY)-(X, Y)
     Line1.Visible = False
     Case "Ellipse":
     Call Ellipse(picDraw.hdc, OldX, OldY, X, Y)
     picDraw.Refresh
     Case "Rectangle":
     picDraw.Line (OldX, OldY)-(X, Y), , B
     Rect1.Visible = False
     Case "FilledRectangle":
     picDraw.Line (OldX, OldY)-(X, Y), , BF
     Rect1.Visible = False
     Case "Dropper"
     shpLineColor.FillColor = picDraw.Point(X, Y)
     picDraw.ForeColor = picDraw.Point(X, Y)
     blnChangeColor = False
     Case "Text"
```

```
        CommonDialog1.Flags = cdlCFScreenFonts
        CommonDialog1.ShowFont
        On Error GoTo ErrorHandler 'If user cancels Common Dialog
        picDraw.FontName = CommonDialog1.FontName
        picDraw.FontSize = CommonDialog1.FontSize
        picDraw.FontBold = CommonDialog1.FontBold
        picDraw.FontItalic = CommonDialog1.FontItalic
        picDraw.FontUnderline = CommonDialog1.FontUnderline
        picDraw.FontStrikethru = CommonDialog1.FontStrikethru
        picDraw.CurrentX = X
        picDraw.CurrentY = Y
        picDraw.Print InputBox("Please Enter Your Text", _
        "Text Tool")
    ErrorHandler:
     End Select
    End Sub

    Private Sub picToolbar_MouseDown(Button As Integer, Shift As Integer, X
    As Single, Y As Single)
     ReleaseCapture
     SendMessage picToolbar.hWnd, WM_NCLBUTTONDOWN, HTCAPTION, 0&
    End Sub

    Private Sub picColor_MouseDown(Button As Integer, Shift As Integer, X
    As Single, Y As Single)
    On Error Resume Next
     shpLineColor.FillColor = picColor.Point(X, Y)
     picDraw.ForeColor = picColor.Point(X, Y)
     blnChangeColor = True
    End Sub

    Private Sub picColor_MouseMove(Button As Integer, Shift As Integer, X
    As Single, Y As Single)
    On Error Resume Next If blnChangeColor Then
     shpLineColor.FillColor = picColor.Point(X, Y) picDraw.ForeColor =
    picColor.Point(X, Y) End If
    End Sub

    Private Sub picColor_MouseUp(Button As Integer, Shift As Integer, X As
    Single, Y As Single)
     blnChangeColor = False
    End Sub
```

```
Private Sub picToolbar2_MouseDown(Button As Integer, Shift As Integer, _
X As Single, Y As Single)
 ReleaseCapture
 SendMessage picToolbar2.hWnd, WM_NCLBUTTONDOWN, HTCAPTION, 0&
End Sub
```

CHAPTER REVIEW

In this chapter, we put together a complete Visual Basic painting application. The greatest portion of code in this chapter was a review of items we previously discussed in an earlier chapter. A few new items were covered in the chapter, the most notable of which is the API calls that allow you to move controls at runtime. We used this feature to make our toolbar and canvas have the ability to be placed anywhere on the screen. Lastly, you learned how to create Undo functionality for this painting application, which can be altered to work in many types of programs.

15 Screen Capture

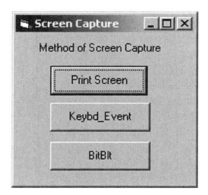

In this chapter, we will develop a Visual Basic application that will function as a screen capture program. You could add these functions to the Paint program from the previous chapter or leave them as a standalone program like the one we are developing.

THE PROGRAM

There are several approaches that we could use to develop this program. The easiest way would be to prompt the end user to press the Print Screen key on their keyboard and retrieve the resulting bitmap from the Windows Clipboard. A second choice would be to use an API function called Keybd_Event. The third option would be to use another API function called BitBlt (Bit Block Transfers). Finally, we could use OLE (Object Linking and Embedding) to handle the screen capture.

The OLE version would work only in Windows 95/98/ME and will not work in Windows 2000 so we don't want to use that route. Other than that particular feature, the others could work to varying degrees of difficulty and options. Therefore, we'll build an application that uses all of these options and allows you to choose the method you would prefer to use.

Print Screen Key

Windows provides the ability to print the entire screen by using the Print Screen key. To try out this function outside of Visual Basic, press the Print Screen button on your keyboard, which is most likely directly to the right of the F12 key on the keyboard. Next, open your favorite drawing program (or Microsoft Paint will do) and choose Edit|Paste (most programs allow you to press CTRL-V as a quick shortcut) on your keyboard. If everything is working, you should see a picture of your entire screen in the program.

While the Print Screen key by itself will only capture the entire screen, there is a simple way to capture only the current active window. To do so, hold down the ALT key on your keyboard and then press the Print Screen key. This will place an image of the active window on the Clipboard and again, you can paste it into a Graphics Editor or even into a word processor like Microsoft Word.

Retrieve Image from Clipboard in Visual Basic

To begin the program, create a Standard EXE application in Visual Basic. Next, create three new forms, then use the Project Explorer and the Properties window to alter the forms so that you have four forms with the following properties:

Name	Caption
frmMain	Screen Capture
frmPrintScrn	Print Screen
frmKeybdEvent	Keybd_Event
frmBitBlt	BitBlt

On frmMain, add three Command Buttons with the following properties:

Name	Caption
cmdPrintScrn	Print Screen
cmdKeybdEvent	Keybd_Event
cmdBitBlt	BitBlt

To finish the GUI for frmMain, we need to add a Label control above the Command Buttons with the Caption property set to "Method of Screen Capture". Move and resize controls until the form appears similar to Figure 15.1. This form will serve as a sort of switchboard to open the other forms.

You need to set frmMain as the Startup Object in the Project Properties window, which can be seen in Figure 15.2. The Startup Object defaults to the form that is created first.

FIGURE 15.1 The final GUI for frmMain.

FIGURE 15.2 The Project Properties window with frmMain as the startup object.

Now that we have the GUI for frmMain finished, double-click the PrintScreen button to open the Code window. The first thing we need to do is clear the Clipboard to erase any old images or text that might be residing on it. Next, we need to display a Message Box informing the user that they need to press the Print Screen button before they continue. Lastly, we'll check the content of the Clipboard to see if it is a bitmap. If so, we will open frmPrintScrn; otherwise, we simply exit the procedure.

The following code can be added:

```
Private Sub cmdPrintScrn_Click()
 Clipboard.Clear
 MsgBox "Please press the Print Screen button _
before clicking OK", vbOKOnly,_
 "VB Print Screen"
 If Clipboard.GetFormat(2) = True Then
 frmPrintScrn.Show
 Else
 End If
End Sub
```

Once frmPrintScrn is displayed, we can simply retrieve the contents of the Clipboard and display it on the form itself. We could create an Image Box or Picture Box, but because we are not planning on doing anything with the image, it's easier to display it on the form. The following code can be added to frmPrintScrn:

```
Private Sub Form_Load()
 Me.Picture = Clipboard.GetData(vbCFBitmap)
End Sub
```

You can now run the program. Close all forms except frmPrintScrn (click the View Object button in the Project Explorer to switch from the Code window to the Form window), then click the Start button on the toolbar to run the program. Click the Print Screen button on the Screen Capture form that appears. Figure 15.3 displays the form and message as they should appear at this point. To finish checking the program, click the Print Screen button on your keyboard and click OK on the message. The screen capture should now appear. Click the End button on the toolbar to stop the program.

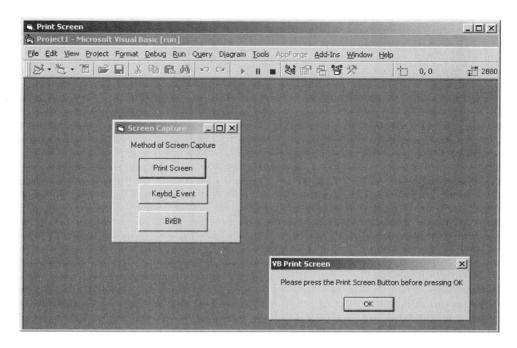

FIGURE 15.3 The Print Screen program is working correctly.

Keybd_Event API

Before we work on the code for the API function, we need to add the code to frmMain that will display frmKeybdEvent when the Keybd_Event button, which is seen in Figure 15.4, is clicked. Unlike the Print Screen example, we don't need to check for anything with this sample. Here is the code you need to add to frmMain:

FIGURE 15.4 frmMain is responsible for displaying the rest of the forms in our program.

```
Private Sub cmdKeybdEvent_Click()
  frmKeybdEvent.Show
End Sub
```

Now that the correct form is displayed when the Keybd_Event button is clicked, we can begin to look at keybd_event. Like any API function, we must first declare it for use in the application. Open frmKeybdEvent and in the General Declarations section of the Code window, add the following:

```
Option Explicit
Private Declare Sub keybd_event Lib "user32.dll" (ByVal bVk As Byte, ByVal bScan As Byte, ByVal dwFlags As Long, ByVal dwExtraInfo As Long)
Const VK_SNAPSHOT As Byte = 44
```

If you would prefer, you can look this up with the API Text Viewer that ships with Visual Basic. We covered this in an earlier chapter, so I won't go over it here. Along with the keybd_event declaration, you probably noticed that we created a constant called VK_SNAPSHOT. It is a virtual-key constant that is used by Visual Basic to simulate the key press we are after. A list of the most common virtual-key codes is located in Appendix D of the book.

The next step is to use the Form_Load event to take the screen shot using the API. If you look at the keybd_event declaration, you will notice that we need to pass four parameters to it. The parameters and their meaning are as follows:

Parameter	Meaning
BYTE bVk	Virtual-key code of simulated key
BYTE bScan	Determines if active screen (0) or full screen (1) are captured
DWORD dwFlags	Optional flags to determine, for example, if key is pressed or released
DWORD dwExtraInfo	Additional value (not used in our example)

The code for the Form_Load event for frmKeybd_Event is very simple. First, we need to clear the contents of the Clipboard to erase any information that is currently residing in it. Next, we will call keybd_event with any optional parameters that we might need. For our example, we'll use the constant VK_SNAPSHOT for the BYTE bVk value and the rest of the parameters will be set to 0. This will capture only the active window, which should result in an image similar to Figure 15.5. Lastly, we use DoEvents to give the program time to update the Clipboard before setting the Picture property of the form equal to the newly acquired image.

The following code can be added to the frmKeybdEvent Code window:

```
Private Sub Form_Load()
 Clipboard.Clear
 Call keybd_event(VK_SNAPSHOT, 0, 0, 0)
 DoEvents
 Me.Picture = Clipboard.GetData(vbCFBitmap)
End Sub
```

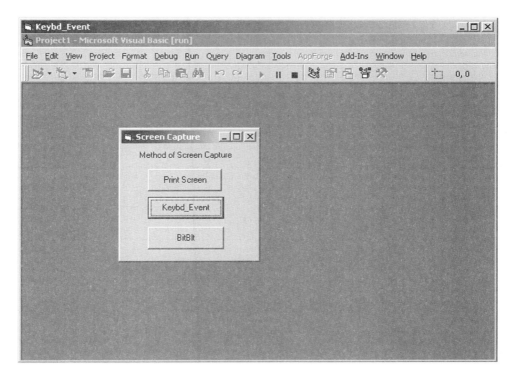

FIGURE 15.5 The active screen is captured using the keybd_event API.

BitBlt

The last solution we'll add to our program is the BitBlt (bit block transfers) option. It will use the BitBlt and several additional API functions to capture either the active window or the full screen. We need to display frmBitBlt when the BitBlt Command Button is clicked on frmMain. Here is the code for this on frmMain:

```
Private Sub cmdBitBlt_Click()
 frmBitBlt.Show
End Sub
```

We need to add two Command Buttons and a Picture Box to frmBitBlt so that they resemble Figure 15.6. The following properties should be set for the controls:

Type	Name	Caption
Command Button	cmdDesktop	Capture Full Screen
Command Button	cmdActiveWindow	Active Window
Picture Box	picScreen	

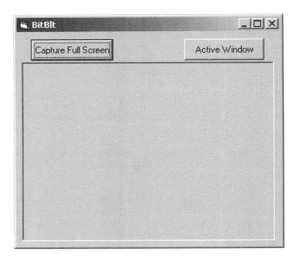

FIGURE 15.6 The GUI for frmBitBlt.

The next step is to create the API declarations. Add these to the Code window of frmBitBlt then we'll look at the functions individually:

```
Option Explicit
Private Declare Function GetWindowRect Lib "user32" (ByVal hwnd As
Long, lpRect As RECT) As Long
Private Declare Function GetDesktopWindow Lib "user32" () As Long
Private Declare Function GetActiveWindow Lib "user32" () As Long
Private Declare Function GetWindowDC Lib "user32" (ByVal hwnd As Long)
As Long
```

```
Private Declare Function BitBlt Lib "gdi32" (ByVal hDestDC As Long,
ByVal x As Long, ByVal y As Long, ByVal nWidth As Long, ByVal nHeight
As Long, ByVal hSrcDC As Long, ByVal xSrc As Long, ByVal ySrc As Long,
ByVal dwRop As Long) As Long
Private Declare Function ReleaseDC Lib "user32" (ByVal hwnd As Long,
ByVal hdc As Long) As Long

Private Type RECT
  Left As Long
  Top As Long
  Right As Long
  Bottom As Long
End Type:
```

Function	Use
GetWindowRect	It is used to read the size and position of a window.
GetDesktopWindow	Returns a handle to the desktop window.
GetActiveWindow	Returns a handle to the current active window.
GetWindowDC	Returns the device context (DC) of a window.
BitBlt	Is used to perform bit-block transfers of a rectangular portion of an image from one device to another.
ReleaseDC	Releases device context.

The next step in the development is to create a GetScreen procedure that will be passed a Windows handle which in turn will be used to capture the active window or the entire desktop depending on the handle being passed. The first step in the procedure is to declare the variables. The variables include the left, right, top, and bottom positions of a rectangle (the window positions), a rectangle and a window device context.

Once the variables have been created, we need to clear picScreen of any current information. Here is the procedure at this point in time for frmBitBlt:

```
Public Sub GetScreen(lWindowhWnd As Long)
  Dim lWindowhDC As Long
  Dim nLeft As Long
  Dim nTop As Long
  Dim nWidth As Long
```

```
Dim nHeight As Long
Dim rRect As RECT

picScreen.Cls
```

The next steps use the API functions to get the window and will set the picScreen width and height accordingly. Lastly, we use BitBlt to transfer the image to picScreen and then release the device context of the window.

Here is the last part of the procedure:

```
GetWindowRect lWindowhWnd, rRect
lWindowhDC = GetWindowDC(lWindowhWnd)
nLeft = 0
nTop = 0
nWidth = rRect.Right - rRect.Left
nHeight = rRect.Bottom - rRect.Top
picScreen.Width = nWidth * Screen.TwipsPerPixelX
picScreen.Height = nHeight * Screen.TwipsPerPixelY
BitBlt picScreen.hdc, 0, 0, nWidth, nHeight, lWindowhDC, nLeft, nTop, vbSrcCopy
ReleaseDC lWindowhWnd, lWindowhDC
End Sub
```

The last step in this part of the program is to add code to the cmdDesktop_Click event and cmdActiveWindow_Click event of frmBitBlt. These procedures are nearly identical, with the only change being the value we pass to the GetScreen procedure; i.e., the active window or desktop window. Here is the code for the two events:

```
Private Sub cmdDesktop_Click()
Dim lhWnd As Long
 lhWnd = GetDesktopWindow
 GetScreen lhWnd
End Sub

Private Sub cmdActiveWindow_Click()
Dim lhWnd As Long
 lhWnd = GetActiveWindow
 GetScreen lhWnd
End Sub
```

Open all the forms, then run the program and click the BitBlt button. When you have finished, you can test the final application by running it. Then, from the Screen Capture window that appears, click each of the three buttons. You've already tested the Print Screen option, so we won't look at it again. The second button, Keybd_Event, should cause a window to be displayed that immediately has captured the Screen Capture window. The final button, BitBlt, displays a window that has two options: Capture Full Screen or Active Window. If you choose Capture Full Screen, you will see the entire desktop captured within the Picture Box. On the other hand, if you click Active Window, you will only see the Active Window displayed in the Picture Box. Figure 15.7 displays all of the possible combinations available for screen capture in this program.

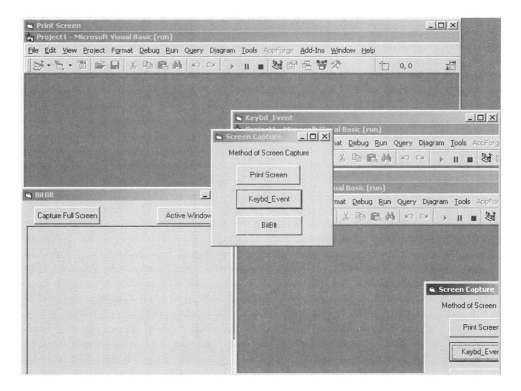

FIGURE 15.7 The final program with all windows being displayed.

COMPLETE CODE LISTING

The following code is the complete listing for this chapter:

```
frmBitBlt:
Option Explicit

Private Declare Function GetWindowRect Lib "user32"_
 (ByVal hwnd As Long, lpRect As RECT) As Long
Private Declare Function GetDesktopWindow Lib "user32" () As Long
Private Declare Function GetActiveWindow Lib "user32" () As Long
Private Declare Function GetWindowDC Lib "user32" _
(ByVal hwnd As Long) As Long
Private Declare Function BitBlt Lib "gdi32" _
(ByVal hDestDC As Long, ByVal x As Long, ByVal y As Long,_
 ByVal nWidth As Long, ByVal nHeight As Long, _
ByVal hSrcDC As Long, ByVal xSrc As Long, ByVal ySrc As Long,_
 ByVal dwRop As Long) As Long

Private Declare Function ReleaseDC Lib "user32" (ByVal hwnd As Long,
ByVal hdc As Long) As Long

Private Type RECT
 Left As Long
 Top As Long
 Right As Long
 Bottom As Long
End Type

Public Sub GetScreen(lWindowhWnd As Long)
 Dim lWindowhDC As Long
 Dim nLeft As Long
 Dim nTop As Long
 Dim nWidth As Long
 Dim nHeight As Long
 Dim rRect As RECT

 picScreen.Cls
 GetWindowRect lWindowhWnd, rRect
 lWindowhDC = GetWindowDC(lWindowhWnd)
 nLeft = 0
 nTop = 0
 nWidth = rRect.Right - rRect.Left
 nHeight = rRect.Bottom - rRect.Top
```

```
  picScreen.Width = nWidth * Screen.TwipsPerPixelX
  picScreen.Height = nHeight * Screen.TwipsPerPixelY
  BitBlt picScreen.hdc, 0, 0, nWidth, nHeight, lWindowDC,_
  nLeft, nTop, vbSrcCopy 'blt to screen
  ReleaseDC lWindowhWnd, lWindowDC
End Sub

Private Sub cmdDesktop_Click()
Dim lhWnd As Long
 lhWnd = GetDesktopWindow
 GetScreen lhWnd
End Sub

Private Sub cmdActiveWindow_Click()
Dim lhWnd As Long
 lhWnd = GetActiveWindow
 GetScreen lhWnd
End Sub
```

frmKeybdEvent:

```
Option Explicit
Private Declare Sub keybd_event Lib "user32.dll" _
(ByVal bVk As Byte, ByVal bScan As Byte, ByVal dwFlags As Long, _
ByVal dwExtraInfo As Long)
Const VK_SNAPSHOT As Byte = 44


Private Sub Form_Load()
 Clipboard.Clear
 Call keybd_event(VK_SNAPSHOT, 0, 0, 0)
 DoEvents
 Me.Picture = Clipboard.GetData(vbCFBitmap)
End Sub
```

frmMain:

```
Private Sub cmdBitBlt_Click()
 frmBitBlt.Show
End Sub
```

```
Private Sub cmdKeybdEvent_Click()
 frmKeybdEvent.Show
End Sub

Private Sub cmdPrintScrn_Click()
 Clipboard.Clear
 MsgBox "Please press the Print Screen button before _
clicking OK", vbOKOnly, "VB Print Screen"
 If Clipboard.GetFormat(2) = True Then
 frmPrintScrn.Show
 Else
 End If
End Sub
```

frmPrintScrn:

```
Private Sub Form_Load()
 Me.Picture = Clipboard.GetData(vbCFBitmap)
End Sub
```

CHAPTER REVIEW

In this chapter we created a screen shot application using a variety of methods including Print Screen and API functions like keybd_event and BitBlt. Although we used API functions in previous chapters, the use of the keybd_event and BitBlt were introduced in this one. BitBlt is useful for a wide range of needs and was the primary way to develop high-speed VB games before DirectX became readily available.

16 Graphing Calculator

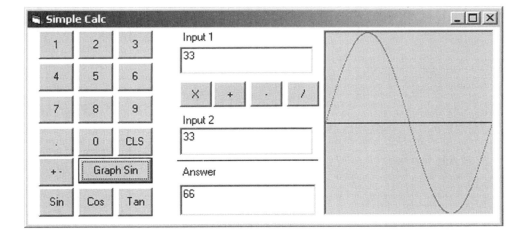

It's very easy to build math-related programs in Visual Basic; in this chapter, we'll explore several built-in functions for math, including some basic trigonometry functions such as Sin, Cos, and Tan. We'll also look at how to use a Picture Box to display graphs of the functions.

THE PROJECT

Creating a GUI

There are several ways we could write the code for the simple calculator that appears in Figure 16.1. We could create several Command Buttons that when clicked could change the value in a Text Box. The problem with this approach is the large amount of code that would be required in the event raised by every button. A better prospect would be to create an array of Command Buttons and utilize the index of the button to determine which button was being clicked.

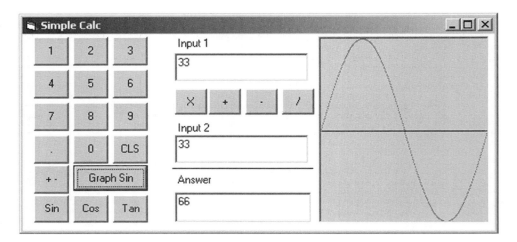

FIGURE 16.1 The calculator that we develop in this chapter will look similar to this one.

You can begin this project by creating a Standard EXE project in Visual Basic. The form should be set to the following properties with the Properties window:

Name: frmCalculator
Caption: Simple Calc
BackColor: ToolTip
Width: 7650
Height: 3660

The next step is to create an array of Command Buttons. You should begin by creating the first button and placing it on the form. It should look similar to Figure 16.2. This button will then be copied and pasted several times to create each of the buttons. Before proceeding with the next step, you should set the Command Button properties as follows:

Name:	cmdNumber
Caption:	1
Height:	465
Width:	555

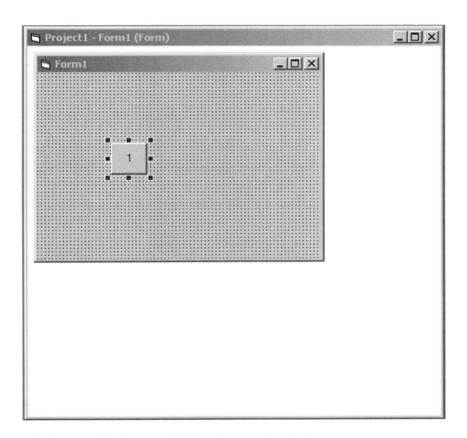

FIGURE 16.2 The Command Button is placed on the form.

The next step is to copy and paste the button several times onto the form. You can do this by clicking on the button and then pressing the CTRL key and the C key simultaneously. You can then use the paste shortcut of CTRL-V to paste it. Otherwise, you could select the button with the mouse and right-click on it. From the pop-up menu that appears (it can be seen in Figure 16.3), you can select Copy. Then you can click on the Edit menu and select Paste.

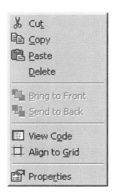

FIGURE 16.3 The pop-up menu should look similar to this figure.

Regardless of the approach to copying, you should see a message similar to Figure 16.4 when you paste it onto the form. The message is asking if you would like to create a control array, which is exactly what we want to do, so you should click the Yes button.

FIGURE 16.4 The control array is created when you click Yes at this prompt.

You can continue pasting the buttons until you have enough for buttons 1 through 9 and 0. You will then need to individually set their Caption properties as follows:

Button (Index)	Caption
cmdButton(0)	1
cmdButton(1)	2
cmdButton(2)	3
cmdButton(3)	4

Button (Index)	Caption
cmdButton(4)	5
cmdButton(5)	6
cmdButton(6)	7
cmdButton(7)	8
cmdButton(8)	9
cmdButton(9)	0

The buttons should then be arranged to appear as Figure 16.5.

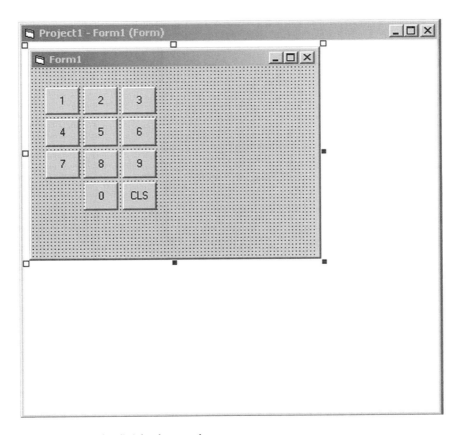

FIGURE 16.5 The finished control array.

Finishing the GUI

There are only a few additional Command Buttons, Text Boxes, Labels and a Picture Box that will round out the GUI for our application. To begin, create the Command Buttons and set their properties as follows:

Name	Caption
cmdMultiply	X
cmdAdd	+
cmdSubtract	-
cmdDivide	/
cmdClear	CLS
cmdDecimal	.
cmdPlusNeg	+ -
cmdGraphSin	Graph Sin
cmdSin	Sin
cmdCos	Cos
cmdTan	Tan

Their widths and heights should be the same as the earlier numbered buttons and they should be placed at locations that are similar to Figure 16.6.

The next step is to place the three Text Boxes on the form. Again, the exact placement of these is unimportant, but they should placed in a manner that resembles Figure 16.7.

Their properties can be set as follows:

Name	Text Property	Width	Height
txtBoxA	blank	2130	465
txtBoxB	blank	2130	465
txtAnswer	blank	2130	465

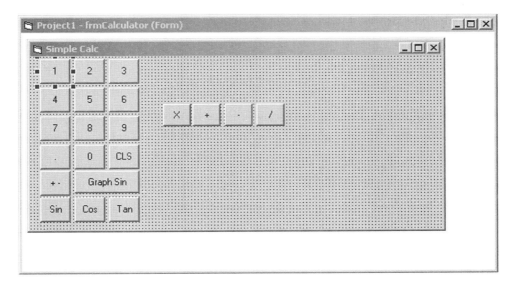

FIGURE 16.6 The additional buttons on the form.

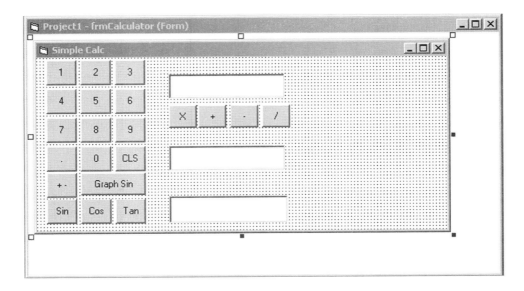

FIGURE 16.7 Text Boxes placed on the GUI.

We also need to create labels for these boxes so that the user of the program will understand what they are for. Create labels with the following properties and place them such that they are in a location similar to Figure 16.8:

Name	Caption	BackColor
Label1	Answer	&H80000018& (ToolTip)
Label2	Input #1	&H80000018& (ToolTip)
Label3	Input #2	&H80000018& (ToolTip)

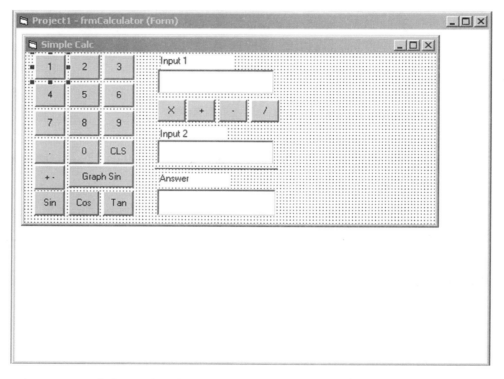

FIGURE 16.8 Labels are placed next to their respective Text Boxes.

The final step that is needed to finish the GUI is the creation of a Picture Box for the displaying of the graph of the Sin function. The Picture Box should be set with the following properties and the final GUI should look something like Figure 16.9.

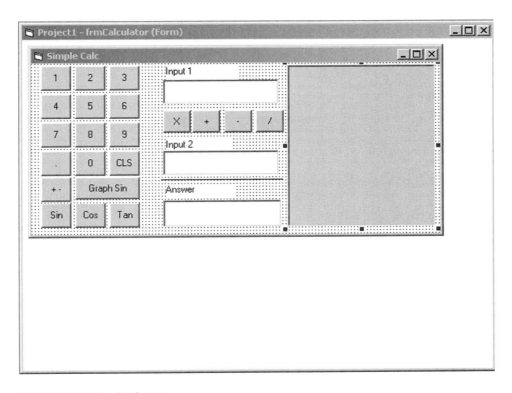

FIGURE 16.9 The final GUI.

WRITING THE CODE

The Command Button Array

With the GUI out of the way, we can concentrate on the code necessary to complete the project. We'll begin by looking at the event that is raised when a button is clicked within the control array. The event is actually very similar to a standard Click event with the addition of an integer value called Index that is used to determine which button was clicked.

```
Private Sub cmdNumber_Click(Index As Integer)

End Sub
```

Before we can use the index to determine the value, we must first determine if the value is supposed to be placed in txtBoxA or txtBoxB. To do this, we need to create a String variable that will hold a character so that we can check to see which box currently needs the input. First, create the following in the General Declarations section of the Code window:

```
Option Explicit
Dim Box As String
```

The next step is to use the GotFocus procedures of the two Text Boxes to change the value of the Box variable. The following code is all that is necessary:

```
Private Sub txtBoxA_GotFocus()
  Box = "A"
End Sub

Private Sub txtBoxB_GotFocus()
  Box = "B"
End Sub
```

Now, we can use the Box variable to ascertain how to add the button value to the appropriate Text Box. To begin, we check this value to determine if it is equal to "A" and if so, we add it to txtBoxA; otherwise, we add it to txtBoxB. Before we write the code, we must take into account that the Index value of 0 corresponds to the number 1. Therefore, we must add 1 to the Index to get the correct value. The following code will take care of this (you can enter this in the procedure where we created it earlier):

```
Private Sub cmdNumber_Click(Index As Integer)
  If Box = "A" Then
  txtBoxA.Text = txtBoxA & (Index + 1)
  Else
  txtBoxB.Text = txtBoxB & (Index + 1)
  End If
End Sub
```

If you were to run the program at this time, you would see that the values are being placed in the appropriate boxes. However, if you click on the button with 0 as a caption, you will see the number 10 being added to a Text Box. We need to add a line to the procedure that checks to see if the Index value is equal to 9 and if so change it to –1. Then, when 1 is added to it, it will equal 0. The line needs to be added to the first line in the procedure, which now reads as follows:

```
Private Sub cmdNumber_Click(Index As Integer)
If Index = 9 Then Index = -1 ' Handle 0
 If Box = "A" Then
 txtBoxA.Text = txtBoxA & (Index + 1)
 Else
 txtBoxB.Text = txtBoxB & (Index + 1)
 End If
End Sub
```

Negative or Positive

Now that the values are being placed in the appropriate Text Boxes, we need to add the positive or negative toggle function that is associated with cmdPlusNeg. We can again use the same check of the Box variable to decide which Text Box is being changed and simply assign it to a negative of itself, thereby adding the toggle function to the program. Here is the code:

```
Private Sub cmdPlusNeg_Click()
 If Box = "A" Then
 txtBoxA.Text = -(txtBoxA)
 Else
 txtBoxB.Text = -(txtBoxB)
 End If
End Sub
```

Trig Functions

The next step is to add functions that will enable the calculation of Sin, Cos and Tan. The value that is generated by these functions will be placed in txtAnswer but again, we must first determine which Text Box (txtBoxA or txtBoxB) that this value should be calculated from. Like the previous steps, we'll use the Box variable. The Sin, Cos and Tan functions are built-in Visual Basic functions that return a double. We can use these functions to create the following procedures:

```
Private Sub cmdSin_Click()
 If Box = "A" Then
 txtAnswer.Text = Sin(txtBoxA)
 Else
 txtAnswer.Text = Sin(txtBoxB)
 End If
End Sub
```

```
Private Sub cmdTan_Click()
 If Box = "A" Then
 txtAnswer.Text = Tan(txtBoxA)
 Else
 txtAnswer.Text = Tan(txtBoxB)
 End If
End Sub

Private Sub cmdCos_Click()
 If Box = "A" Then
 txtAnswer.Text = Cos(txtBoxA)
 Else
 txtAnswer.Text = Cos(txtBoxB)
 End If
End Sub
```

Decimal Points

The decimal point might be the easiest item to deal with in the entire program. You can simply add a "." to the currently selected Text Box. The following code is all that is necessary for this step:

```
Private Sub cmdDecimal_Click()
 If Box = "A" Then
 txtBoxA.Text = txtBoxA & "."
 Else
 txtBoxB.Text = txtBoxB & "."
 End If
End Sub
```

Add, Subtract, Multiply, and Divide

To add, subtract multiply, or divide the numbers in txtBoxA and txtBoxB, we first need to check to see if they have a number in them. Otherwise, the calculations will cause several errors. To do this, we can simply check to see if we have something in the Text Box. If we don't, then we can do the calculation; otherwise, we exit the procedure.

The calculated answers are stored and displayed in txtAnswer. Because we have two input boxes, we can simply subtract, multiply or divide with txtBoxA and txtBoxB. The only mathematical option that can cause a problem is addition, which needs to use the Val keyword, which returns the numbers contained in a string as a numeric value. This allows us to add the numbers instead of combining the strings to form a new string.

The procedures for adding, subtracting, multiplying and dividing are complete as follows:

```
Private Sub cmdMultiply_Click()
  If txtBoxA <> "" And txtBoxB <> "" Then
  txtAnswer = txtBoxA * txtBoxB
  End If
End Sub

Private Sub cmdSubtract_Click()
  If txtBoxA <> "" And txtBoxB <> "" Then
  txtAnswer = txtBoxB - txtBoxA
  End If
End Sub

Private Sub cmdDivide_Click()
  If txtBoxA <> "" And txtBoxB <> "" And txtBoxB <> "0" Then
  txtAnswer = txtBoxA / txtBoxB
  End If
End Sub

Private Sub cmdAdd_Click()
  If txtBoxA <> "" And txtBoxB <> "" Then
  txtAnswer = Val(txtBoxA) + Val(txtBoxB)
End Sub
```

GRAPHING TRIG FUNCTIONS

In order to graph the Trig functions, we need something to output the data to. We could use a form itself, but in our case, because it is full of controls, we'll use the picGraph Picture Box that we have already placed on the form. The Picture Box can be seen in Figure 16.10.

FIGURE 16.10 The Picture Box will display the graph.

The first thing we need to do is set up the Picture Box to a specific scale. We're planning on graphing the Sin function so we'll set the scale width to 360 with a value of 1 at the upper left and a −1 at the lower right. This will allow us to use the entire box to display the graph. Picture Boxes have a Scale property that we'll use to accomplish this task.

The next step will be to draw a straight line from 0,0 to 360,0. This line is not any part of the Sin function, but will serve as a means to determine values on the graph. After the line is drawn, we can move to the next step, which is converting the Degree values to radians. If you are unfamiliar with the formula, you can convert degrees to radians by taking the degree value and multiplying it by π and dividing by 180. Visual Basic does not have any built-in value for π, therefore, we will create a constant for π giving it a value of 3.1415. For more accurate calculations, you could use several more decimal places, but for this program, this is already more than we need.

Next, we'll create a For…Next loop that will serve to step through every degree from 0 to 360. Inside this loop we will place the conversion from degrees to radians and then graph the Sin function using the results from the calculation along with the value of the For loop counter as the degree. We can use the Picture Box PSet, which creates a point at a specific x and y location, to accomplish the task.

You can add the following code to the program:

```
Private Sub cmdGraphSin_Click()
Const pi As Double = 3.1415
  Dim intDegrees As Integer
  Dim sngRadians As Single
  
  picGraph.Scale (0, 1)-(360, -1)
  picGraph.Line (0, 0)-(360, 0)
  For intDegrees = 0 To 360
    sngRadians = intDegrees * pi / 180
    picGraph.PSet (intDegrees, Sin(sngRadians)), vbRed
  Next intDegrees
End Sub
```

FINAL STEPS

The final steps in the program are to create the procedures that will clear the Text Boxes and the Form_Load event that will reset the initial values of the program. Here are the finished procedures, which can be added to the program:

```
Private Sub cmdClear_Click()
  txtBoxA = ""
  txtBoxB = ""
End Sub

Private Sub Form_Load()
  Box = "A"
  txtBoxA = ""
  txtBoxB = ""
End Sub
```

If you run the application at this point, you can test it by entering 33 into Input 1 and into Input 2. Then, click the "+" button and the Graph Sin button and the screen should look like Figure 16.11.

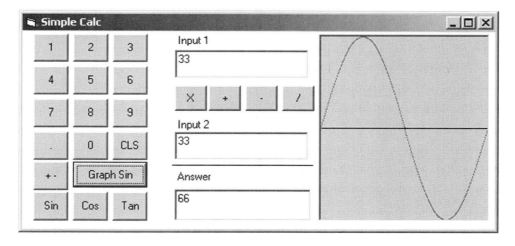

FIGURE 16.11 The completed calculator.

COMPLETE CODE LISTING

The following code is the complete listing for this chapter:

```
Option Explicit
Dim Box As String

Private Sub cmdAdd_Click()
  If txtBoxA <> "" And txtBoxB <> "" Then
```

```vb
   txtAnswer = Val(txtBoxA) + Val(txtBoxB)
  End If
End Sub

Private Sub cmdClear_Click()
 txtBoxA = ""
 txtBoxB = ""
End Sub

Private Sub cmdCos_Click()
 If Box = "A" Then
 txtAnswer.Text = Cos(txtBoxA)
 Else
 txtAnswer.Text = Cos(txtBoxB)
 End If
End Sub

Private Sub cmdDecimal_Click()
 If Box = "A" Then
 txtBoxA.Text = txtBoxA & "."
 Else
 txtBoxB.Text = txtBoxB & "."
 End If
End Sub

Private Sub cmdDivide_Click()
 If txtBoxA <> "" And txtBoxB <> "" And txtBoxB <> "0" Then
 txtAnswer = txtBoxA / txtBoxB
 End If
End Sub

Private Sub cmdGraphSin_Click()
Const pi As Double = 3.1415
 Dim intDegrees As Integer
 Dim sngRadians As Single

 picGraph.Scale (0, 1)-(360, -1)
 picGraph.Line (0, 0)-(360, 0)
 For intDegrees = 0 To 360
 sngRadians = intDegrees * pi / 180
 picGraph.PSet (intDegrees, Sin(sngRadians)), vbRed
 Next intDegrees
End Sub
```

```
Private Sub cmdNumber_Click(Index As Integer)
 If Index = 9 Then Index = -1 ' Handle 0
 If Box = "A" Then
 txtBoxA.Text = txtBoxA & (Index + 1)
 Else
 txtBoxB.Text = txtBoxB & (Index + 1)
 End If
End Sub

Private Sub cmdMultiply_Click()
 If txtBoxA <> "" And txtBoxB <> "" Then
 txtAnswer = txtBoxA * txtBoxB
 End If
End Sub

Private Sub cmdPlusNeg_Click()
 If Box = "A" Then
 txtBoxA.Text = -(txtBoxA)
 Else
 txtBoxB.Text = -(txtBoxB)
 End If
End Sub

Private Sub cmdSin_Click()
 If Box = "A" Then
 txtAnswer.Text = Sin(txtBoxA)
 Else
 txtAnswer.Text = Sin(txtBoxB)
 End If
End Sub

Private Sub cmdSubtract_Click()
 If txtBoxA <> "" And txtBoxB <> "" Then
 txtAnswer = txtBoxB - txtBoxA
 End If
End Sub

Private Sub cmdTan_Click()
 If Box = "A" Then
 txtAnswer.Text = Tan(txtBoxA)
 Else
 txtAnswer.Text = Tan(txtBoxB)
 End If
End Sub
```

```
Private Sub Form_Load()
 Box = "A"
 txtBoxA = ""
 txtBoxB = ""
End Sub

Private Sub txtBoxA_GotFocus()
 Box = "A"
End Sub

Private Sub txtBoxB_GotFocus()
 Box = "B"
End Sub
```

CHAPTER REVIEW

It's very easy to build math-related programs in Visual Basic and in this chapter, we used several of the built-in functions for math, including some basic trigonometry functions such as Sin, Cos, and Tan. You were introduced to the command array that we effectively used for the calculator buttons. Lastly, we used a Picture Box to display graphs of the Sin function.

17 Slot Machine

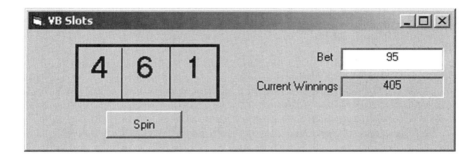

In this chapter we'll develop a Visual Basic Slot Machine using labels to display a series of random numbers. As you can see in Figure 17.1, it will function like a basic slot machine with a betting system and a predetermined amount of initial funds for the user. The random numbers will be checked to determine if at least two of them match. If they do, the player will receive twice the amount they wagered; otherwise, they will lose the money.

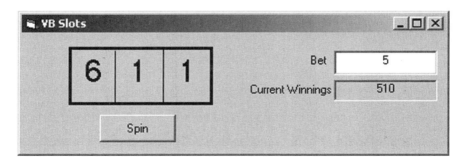

FIGURE 17.1 The Visual Basic Slot Machine.

BEGINNING THE PROJECT

A Basic GUI

Oftentimes, slot machines use a series of pictures as their interface. Instead of this approach, we'll use three Label controls that will display random numbers. The random numbers will be between 1 and 9, and if two of them match on a given spin, the user wins.

To begin, create a new Standard EXE project in Visual Basic. Next, create a Label control with the following properties:

Name	Caption	Width	Height	Font and Size
lblSlot 1		510	735	Sans Serif 24 pt.

Next, copy and paste it, creating a control array. The controls should be placed on the form so that they appear similar to Figure 17.2.

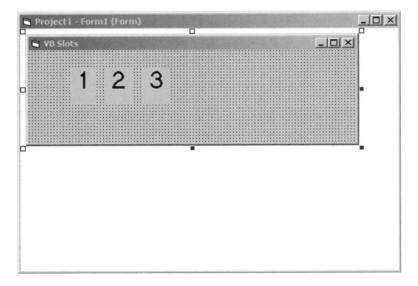

FIGURE 17.2 The Label controls will display the random numbers that determine if you win or lose.

Next, we need to create a Command Button, a Text Box, and an additional Label control with the following properties:

Control	Name	Caption
Command Button	cmdSpin	Spin
Label	lblWin	500
Text Box	txtBet	0

Once the properties have been set on the controls, they can be placed on the form in an arrangement similar to Figure 17.3. You can also create the labels that are displayed next to txtBet and lblWin in Figure 17.4.

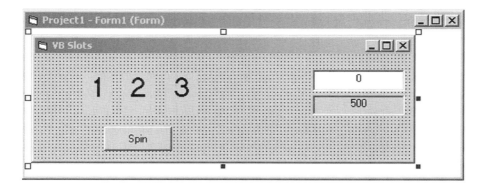

FIGURE 17.3 The GUI is nearly completed.

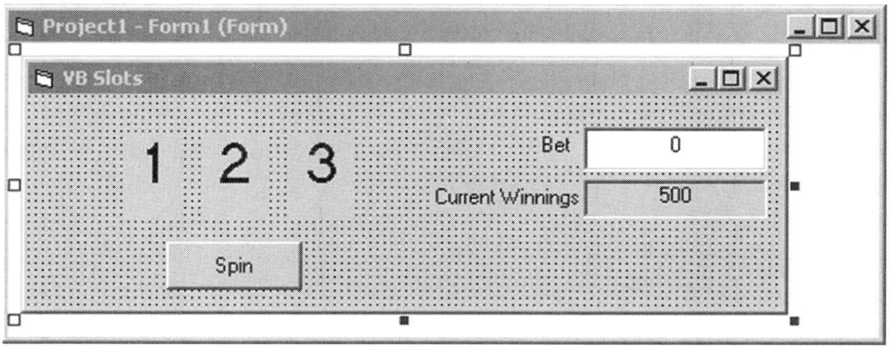

FIGURE 17.4 Labels are needed to describe the functions of the controls.

The last step required to create the GUI is the use of a Shape control and a couple of lines that will separate the Label control array. You can create something similar to Figure 17.5.

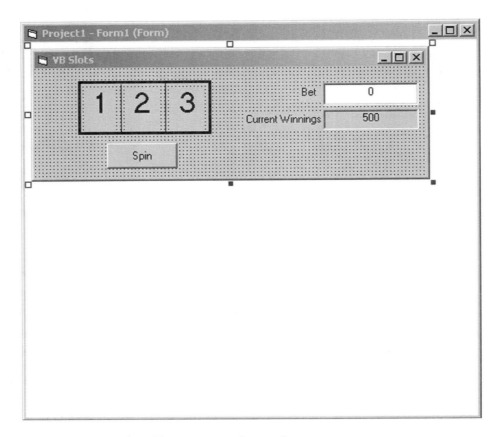

FIGURE 17.5 Shape and lines separate the numbers.

Spinning the Wheel

The first step in writing the code deals with handling the spinning of the slot machine. We have already decided that we will use random numbers that will be assigned to each of the label controls in the lblSlot control array. We need to create the following variable declaration which will step through the control array.

```
Option Explicit
Dim intNumb As Integer
```

Now that we have the declaration out of the way, we can focus on the program functions. First, we will use the cmdSpin_Click event that is raised when the Spin button is clicked. We can begin the cmdSpin_Click procedure by setting the Enabled property of the button to False so that the user cannot click the button again until the spin is completed.

Next, we check the value of txtBet to see if it is less than 1. In our game, the minimum bet is $1, so we will prompt the user with a Message Box, which can be seen in Figure 17.6, if their bet is less than the minimum. If it is less than the minimum, the program will continue by setting the Enabled property of cmdSpin to True and then exit the subprocedure.

```
Private Sub cmdSpin_Click()
  cmdSpin.Enabled = False

  If Val(txtBet) < 1 Then
  MsgBox "You need to bet a minimum of $1", vbOKOnly, "Betting Error"
  cmdSpin.Enabled = True
  Exit Sub
  End If
End Sub
```

If the value of txtBet is greater than or equal to 1 then the program will continue. This is where we will use the intNumb variable that we declared earlier. Using a For…Next loop, we'll step from 0 to 2 and in doing so, we'll set the lblSlot control array, with an index value that is the same as the value of the variable, equal to a random number between 1 and 9.

FIGURE 17.6 A Message Box is used to inform the user that the minimum bet is $1.

We need to delay the program between steps so that it appears that the labels are spinning one at a time and then stop on the final values. Once the three numbers have been created, we need to determine if the user wins or loses. Here is the complete code to the cmdSpin_Click procedure that we started in the previous step:

```
Private Sub cmdSpin_Click()
  cmdSpin.Enabled = False

  If Val(txtBet) < 1 Then
  MsgBox "You need to bet a minimum of $1", vbOKOnly, "Betting Error"
```

```
    cmdSpin.Enabled = True
    Exit Sub
End If

For intNumb = 0 To 2
    Delay 1
    Randomize
    lblSlot(intNumb).Caption = Int(Rnd * 9) + 1
Next
CheckWin
End Sub
```

You may notice the Delay 1 statement and the CheckWin statements that are present in the procedure. They are calling procedures that we will create in the next steps.

Creating a Delay

To create a delay in the program, we can create a simple procedure that uses the built-in Visual Basic time functions. First, create the procedure as follows:

```
Public Sub Delay(Future As Date)

End Sub
```

The procedure will use the Future as Date value, which it will receive when we call the procedure (i.e., Delay 5), to determine how many seconds it will delay the program between spins. Next, we need to declare a variable called tmpTime that will be used to store time in the future that will resume operations. We will use a While…Wend loop to compare the current time with the future time and if the future time is greater than the current time, we can randomly set the lblSlot variable to a random number between 1 and 9. The While…Wend loop executes statements as long as a given condition is true so it's very useful for our application. We can test the value of the future time and once it's less than the current time, the program will resume operations. We also need to add a DoEvents command inside the loop, which will allow Windows to continue to work properly.

```
Dim tmpTime As Date
 tmpTime = DateAdd("s", HowLong, Now)
 While tmpTime > Now
 lblSlot(intNumb).Caption = Int(Rnd * 9) + 1
 DoEvents 'OK to handle other tasks
 Wend
```

When the procedure finishes its delay, it returns operation to the calling procedure (cmdSpin_Click), which then utilizes Randomize to help keep the final random numbers as random as possible.

WIN OR LOSE

If you were to run the program at this time, it would display the GUI properly and the Label control array would be assigned to random values. However, you would never gain or lose money because we haven't added the CheckWin procedure to determine if you win or lose. We can use an If…Then…Or statement to check the values to see if any two of them are alike. If they are equal, we need to add double the amount of the wager to lblWin. Otherwise, we need to subtract the value from lblWin. Lastly, we need to set the Enabled property of cmdSpin to True so that the user can spin again.

Here is the code for the CheckWin procedure:

```
Private Sub CheckWin()
 If lblSlot(0) = lblSlot(1) Or lblSlot(1) = lblSlot(2) Or lblSlot(0) = lblSlot(2) Then
 lblWin.Caption = Val(lblWin.Caption) + (2 * Val(txtBet))
 Else
 lblWin.Caption = Val(lblWin.Caption) - Val(txtBet)
 End If
 cmdSpin.Enabled = True
End Sub
```

The final program should appear similar to Figure 17.7.

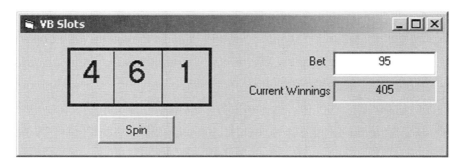

FIGURE 17.7 The finished slot machine.

COMPLETE CODE LISTING

The following code is the complete listing for this chapter:

```
Option Explicit
Dim intNumb As Integer

Public Sub Delay(Future As Date)
 Dim tmpTime As Date
 tmpTime = DateAdd("s", Future, Now)
 While tmpTime > Now
 lblSlot(intNumb).Caption = Int(Rnd * 9) + 1
 DoEvents 'OK to handle other tasks
 Wend
End Sub

Private Sub CheckWin()
 If lblSlot(0) = lblSlot(1) Or lblSlot(1) = lblSlot(2) Or lblSlot(0) = lblSlot(2) Then
 lblWin.Caption = Val(lblWin.Caption) + (2 * Val(txtBet))
 Else
 lblWin.Caption = Val(lblWin.Caption) - Val(txtBet)
 End If
 cmdSpin.Enabled = True
End Sub

Private Sub cmdSpin_Click()
 cmdSpin.Enabled = False

 If Val(txtBet) < 1 Then
 MsgBox "You need to bet a minimum of $1", vbOKOnly, "Betting Error"
 cmdSpin.Enabled = True
 Exit Sub
 End If

 For intNumb = 0 To 2
 Delay 1
 Randomize
 lblSlot(intNumb).Caption = Int(Rnd * 9) + 1
 Next
 CheckWin
End Sub
```

CHAPTER REVIEW

The slot machine uses a number of features from previous applications, including control arrays. It also introduces you to a way that you can safely delay the functions of a program by checking the current time with a time in the future. Lastly, you were introduced to the DoEvents command, which was present in the Time Delay procedure. You can add easily add some functions to the program such as replacing the Labels with Picture Boxes, and adding the ability to save a current game.

18 VB Encryption

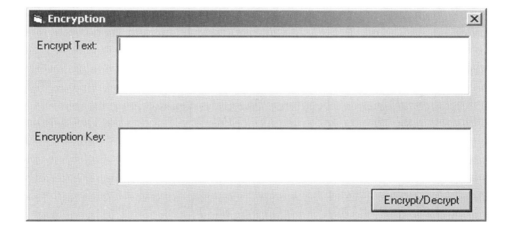

The ability to maintain privacy has always been important and with the proliferation of modern technologies, many feel as though their privacy is under attack. This is especially true when you use electronic information where there is often a need for an efficient method of data encryption. Therefore, depending on the type of application you are developing, it may be important to provide some basic form of encryption for the users of the program.

ENCRYPTION BASICS

There have been volumes of discrete math and computer science books written on the topic of encryption, so obviously this chapter will only provide a very basic introduction to it. One of the easiest forms of encryption was used by Julius Caesar and is called the Caesar Cipher.

In this form of encryption, every letter is replaced with a different letter obtained by a shift in the alphabet. An example is as follows:

Text: A B C D E F G H I J K L M N O P Q R S T U V W X Y Z
Cipher: G H I J K L M N O P Q R S T U V W X Y Z A B C D E F

With this in mind, BOOK would be changed to HUUQ.

The problem with this type of approach is the ability for individuals to decode the messages with a relatively small amount of work. Another easy method is to use the XOR Boolean operator.

XOR

One of the easiest and common methods of encryption (and the one that we will use in this chapter) is the use of the XOR Boolean operator. The XOR Boolean operator is used because it's very easy to understand and the resulting encrypted information can be returned to its original state by simply doing the XOR operation on it again. Depending on your required level of security, the XOR method could be more than sufficient.

For those of you unfamiliar with XOR, we'll look over a little of the basic information. XOR stands for "exclusive or". In other words, it returns a True value if and only if one of the values being compared is True. You may still be confused, and if this is the case, don't worry. We'll look at a few examples.

To begin, look at these examples:

- True XOR True = False
- True XOR False = True
- False XOR True = True
- False XOR False = False
- 1 XOR 0 = 1
- 1 XOR 1 = 0
- 0 XOR 1 = 1
- 0 XOR 0 = 0

Now, look at it when we use a Key value to encrypt a value. A key is simply a series of characters that we use to encrypt information. We can then use the key at a later date to return the value to the original values. Here is an example:

Value	11111111	
	XOR	
Key	00011000	
Encrypted	11100111	

The interesting part of XOR is that we can return the encrypted result to the original value by simply using the XOR operator again:

Encrypted Value	11100111	
	XOR	
Key	00011000	
Original Value	11111111	

CREATING THE APPLICATION

A Quick GUI

The creation of an XOR based encryption program in Visual Basic is a relatively simple process. In our example, we'll have two Text Boxes. The first of the Text Boxes will store our information that we want to encrypt. The second will store a randomly generated key. Lastly, the original Text Box will store the resulting encrypted text information.

We can begin the application with the creation of a GUI, which will consist of two Text Boxes, two labels, and a Command Button. They should be set with the following properties:

Name	Type	Caption	Text
cmdEncrypt	Command Button	Encrypt/Decrypt	
Label1	Label	Encrypt Text:	
Label2	Label	Encryption Key:	
txtText	Text Box		blank
txtKey	Text Box		blank

You should place these items on the form in an arrangement similar to Figure 18.1.

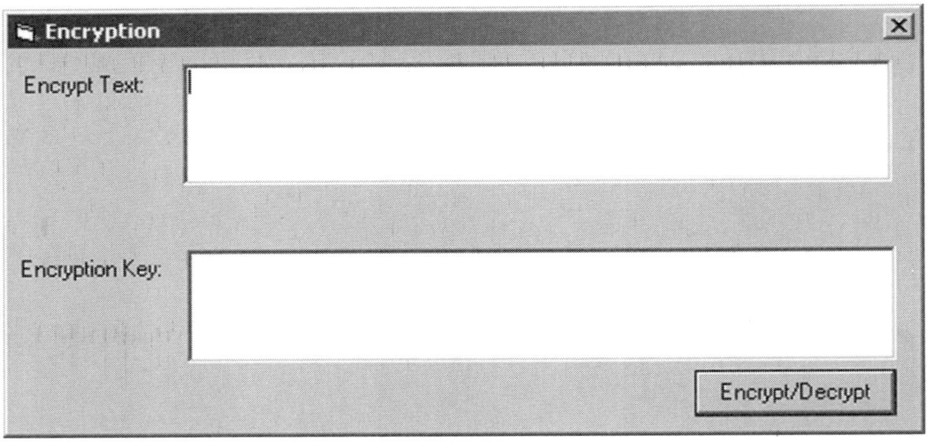

FIGURE 18.1 The final GUI for the encryption program.

THE CODE

Writing the code for the application begins with the creation of the variables that will be used in the program. We need variables for counting through each individual character in the Text Boxes, and a place to store the characters being read from the Text Boxes.

The following list of variables should be added to the program:

```
Option Explicit
Dim intCounter As Integer
Dim strText As String
Dim strKey As String
Dim strChar1 As String * 1
Dim strChar2 As String * 1
Dim intKeyChar As Integer
```

The next step is to use the event that is raised when the Encrypt/Decrypt button is clicked to determine if the txtKey Text Box is empty. If it is empty, we need to create a random key; otherwise, we move on. At this point, let's create the CreateKey subprocedure.

The CreateKey subprocedure will be responsible for the creation of a random set of characters that will be combined to form a random encryption key. To begin, we first set the strText variable equal to txtText. This allows us to check the length of the text that is going to be encrypted and create a key that is of similar length. You could create a smaller or larger encryption key, but for small text boxes, this is probably the easiest and best way to do it.

Our next step is to use the length of the text information to step through the creation of a key based on a series of Random numbers. We'll call the Randomize function at every step, and then append a character based on the random numbers to the end of the txtKey Text Box.

The following procedure is complete:

```
Private Sub CreateKey()
  strText = txtText.Text

  For intCounter = 1 To Len(strText)
  Randomize
  txtKey.Text = txtKey.Text & Chr(Int(Rnd(1) * 96 + 33))
  Next intCounter
End Sub
```

The Click event that was raised in an earlier step will then continue on by setting the strKey variable equal to the txtKey Text Box. We can then count through every character of the strText variable and convert it with XOR. Once we have gone through all of the characters, we then display the results by setting txtText to strTemp.

The following code is the complete subprocedure:

```
Private Sub cmdEncrypt_Click()
Dim strTemp As String

  strText = txtText.Text
  If txtKey = "" Then
  CreateKey
  End If
  strKey = txtKey.Text

  For intCounter = 1 To Len(strText)
  strChar1 = Mid(strText, intCounter, 1)
  intKeyChar = ((intCounter - 1) Mod Len(strKey)) + 1
  strChar2 = Mid(strKey, intKeyChar, 1)
  strTemp = strTemp & Chr(Asc(strChar1) Xor Asc(strChar2))
  Next intCounter
```

```
        txtText.Text = strTemp
    End Sub
```

Again, the great thing about using XOR is that we can simply click on the same button to encrypt/decrypt without changing a single line of code. You could add additional functionality to this program to save the encrypted text and key so that you could later turn the encrypted information back to a regular message. You could also change it so that it would allow you to set the encryption key manually or use a random sequence. If you run the program, it should appear similar to Figure 18.2.

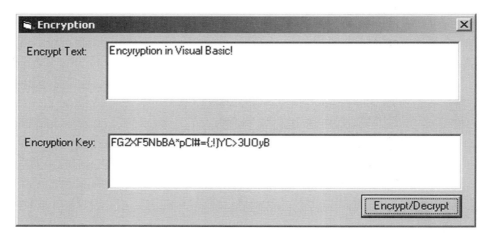

FIGURE 18.2 The completed encryption program.

COMPLETE CODE LISTING

The following code is the complete listing for this chapter:

```
Option Explicit
Dim intCounter As Integer
Dim strText As String
Dim strKey As String
Dim strChar1 As String * 1
Dim strChar2 As String * 1
Dim intKeyChar As Integer

Private Sub CreateKey()
```

```
    strText = txtText.Text

    For intCounter = 1 To Len(strText)
    Randomize
    txtKey.Text = txtKey.Text & Chr(Int(Rnd(1) * 96 + 33))
    Next intCounter
End Sub

Private Sub cmdEncrypt_Click()
Dim strTemp As String

    strText = txtText.Text

    If txtKey = "" Then
    CreateKey
    End If

    strKey = txtKey.Text

    For intCounter = 1 To Len(strText)
    strChar1 = Mid(strText, intCounter, 1)
    intKeyChar = ((intCounter - 1) Mod Len(strKey)) + 1
    strChar2 = Mid(strKey, intKeyChar, 1)
    strTemp = strTemp & Chr(Asc(strChar1) Xor Asc(strChar2))
    Next intCounter

    txtText.Text = strTemp

End Sub
```

CHAPTER REVIEW

Having the ability to keep information private may be important, depending on the type of applications you are going to develop. In this program, we used the XOR method of encryption which is an easy but effective solution to basic encryption needs. You could add features to this program, such as the ability to save the randomly generated key and text information so that they could be reloaded at a later date. You could also save database contents or even pictures in this manner.

Appendix A

STANDARD NAMING CONVENTIONS

While not a requirement, it's a good idea, especially when multiple developers are working on the same project, that you adhere to some type of naming convention for your variables. For instance, the following code might be hard to read if you were unfamiliar with the purpose of the variables:

```
For X = 1 to 10
    Z = X * Y
Next X
```

Instead, you could use something like:

```
For intLength = 1 to 10
    intHeight = intLength * intWidth
Next intLength
```

Visual Basic uses a Scope and Type identifer to prefix each variable name, so it is easy to tell the data type of the variable and where the variable is declared.

The scope is as follows:

Scope	Prefix	Example
Global	g	gstrFirstName
Module-level	m	mblnLastName
Local to Procedure	None	dblFirstName

You may have noticed that the variables have extra information associated with their prefixes. This is because it's important to also name the variables such that anyone glancing at the code is instantly able to determine the type of variable as well. The following list is a sample of the data types and prefixes:

Data Type	Prefix	Example
Boolean	bln	blnFirstTime
Byte	byt	bytData
Collection Object	col	colInformation
Currency	cur	curSales
Date (Time)	dtm	dtmTomorrow
Double	dbl	dblValue
Error	err	errCalculation
Integer	int	intAmount
Long	lng	lngMath
Object	obj	objLast
Single	sng	sngNew
String	str	strEmployeeName
User-defined Type	udt	udtSalesman
Variant	vnt	vntAnything

The common controls also benefit from standard naming:

Control Type	Prefix	Example
3D Panel	pnl	pnlGroup
ADO	Data ADO	adoBiblio
Animated Button	ani	aniClick
Check box	chk	chkDoIt
Combo box	cbo	cboState
Command button	cmd	cmdStart
Common dialog	dlg	dlgSave
Communications	com	comFax
Control	ctr	ctrCurrent
Data	dat	datBiblio
Data-bound Combo Box	dbcbo	dbcboLanguage
Data-bound Grid	dbgrd	dbgrdQueryResult
Data-bound List Box	dblst	dblstJobType
Data Combo	dbc	dbcAuthor
Data Grid	dgd	dgdTitles

Data List	dbl	dblPublisher
Data Repeater	drp	drpLocation
Date Picker	dtp	dtpPublished
Directory list box	dir	dirRoot
Drive list box	drv	drvRoot
File list box	fil	filStatistics
Flat Scroll Bar	fsb	fsbMove
Form	frm	frmStartUp
Frame	fra	fraEmployeeInfo
Gauge	gau	gauToday
Graph	gra	graSales
Grid	grd	grdDates
Hierarchical Flexgrid	flex	flexInventory
Horizontal Scroll Bar	hsb	hsbPosition
Image	img	imgHouse
Image Combo	imgcbo	imgcboItems
Image List	ils	ilsPictures
Label	lbl	lblInformation
List box	lst	lstCount
Lightweight Check Box	lwchk	lwchkArea
Lightweight Combo Box	lwcbo	lwcboState
Lightweight Command Button	lwcmd	lwcmdExit
Lightweight Frame	lwfra	lwfraList
Lightweight Horizontal Scroll Bar	lwhsb	lwhsbAlignment
Lightweight List Box	lwlst	lwlstDetails
Lightweight Option Button	lwopt	lwoptMarried
Lightweight Text Box	lwtxt	lwoptName
Lightweight Vertical Scroll Bar	lwvsb	lwvsbValue
Line	lin	linDesign
List Box	lst	lstData
ListView	lvw	lvwIdeas
MAPI Message	mpm	mpmMessage
MAPI Session	mps	mpsSession

MCI	mci	mciVideo
Menu	mnu	mnuExit
Month View	mvw	mvwDates
MS Chart	ch	chSalesbyRegion
MS Flex Grid	msg	msgFigures
MS Tab	mst	mstFinancials
OLE Container	ole	oleWorksheet
Option button	opt	optSound
Picture box	pic	picMap
Picture Clip	clp	clpToolbar
ProgressBar	prg	prgConvert
RichTextBox	rtf	rtfSales
Remote Data	rd	rdData
Shape	shp	shpLine
Slider	sld	sldValues
Spin	spn	spnInfo
StatusBar	sta	staQuery
SysInfo	sys	sysMy
TabStrip	tab	tabOptions
Text box	txt	txtInformation
Timer	tmr	tmrFirst
Toolbar	tlb	tlbMenu
TreeView	tre	treData
Vertical Scroll Bar	vsb	vsbValue

Appendix B

At this point in time, you should give yourself a big pat on the back. If you've followed through the book, you have created several useful applications that provide a good foundation for the future. Unfortunately, as you learn more about Visual Basic, you'll undoubtedly have a need for answers on specific problems that you will inevitably encounter.

With the large number of Internet resources and Newsgroups available, you can probably find someone willing to help in those times of crisis. The following list will be a great place to start:

Internet Sites:

http://www.vbinformation.com/

http://www.beadsandbaubles.com/coolvb/boards.shtml

http://www.cgvb.com/

http://www.vbhow.to/

http://www.microsoft.com/

Newsgroups:

comp.lang.basic.visual.announce

comp.lang.basic.visual.database

comp.lang.basic.visual.misc

comp.lang.basic.visual.3rdparty

microsoft.public.activex.controls.chatcontrol

microsoft.public.activex.programming.control.webdc

microsoft.public.activex.programming.control.webwiz

microsoft.public.inetexplorer.ie4.activex_contrl

microsoft.public.windows.inetexplorer.ie5.programming.activexcontrol

microsoft.public.vb.3rdparty
microsoft.public.vb.6.webdevelopment
microsoft.public.vb.addins
microsoft.public.vb.bugs
microsoft.public.vb.com
microsoft.public.vb.controls
microsoft.public.vb.controls.creation
microsoft.public.vb.controls.databound
microsoft.public.vb.controls.internet
microsoft.public.vb.crystal
microsoft.public.vb.database
microsoft.public.vb.database.ado
microsoft.public.vb.database.dao
microsoft.public.vb.database.odbc
microsoft.public.vb.database.rdo
microsoft.public.vb.dataenvreport
microsoft.public.vb.deployment
microsoft.public.vb.enterprise
microsoft.public.vb.general.discussion
microsoft.public.vb.installation
microsoft.public.vb.ole
microsoft.public.vb.ole.automation
microsoft.public.vb.ole.cdk
microsoft.public.vb.ole.servers
microsoft.public.vb.setupwiz
microsoft.public.vb.syntax
microsoft.public.vb.vbce
microsoft.public.vb.visual_modeler
microsoft.public.vb.webclasses
microsoft.public.vb.winapi
microsoft.public.vb.winapi.graphics
microsoft.public.vb.winapi.networks

Appendix C

UPGRADING TO VISUAL BASIC.NET

As was mentioned in Chapter 1, Visual Basic.NET will not automatically upgrade an application from Visual Basic 6. It will typically open a VB 6 application and make some conversions, but you will usually be required to modify it manually.

The version of Visual Basic.NET that was available during the writing of this book was a very early Beta version. The vast majority of large or complicated projects that were run through the conversion process were unsuccessful although it's certain that the upgrades that will undoubtedly occur before the final version is released will repair these issues.

New Features

Visual Basic.NET is the next generation of Visual Basic and has been completely reengineered by Microsoft. Visual Basic.NET introduces some changes to the IDE, but it is the addition of Windows Forms and Web Forms that are the most interesting upgrades. The decision to include a wide range of new features made it a necessity to upgrade VB from the ground up and is not just an upgrade to Visual Basic 6.

Upgrading Projects

The upgrade from VB 5 to VB 6 was minor compared with this complete overhaul. If you've been using Visual Basic for awhile, you'll remember the difficulties associated with upgrading from version 3 to 4. Unfortunately, if you had problems with that upgrade, this one will be even harder to manage and upgrading your projects will be much more difficult. For instance, Visual Basic is now a true object-oriented programming language and some features have been removed from the language. Functions such as GoSub…Return have been completely removed.

Because of the new changes associated with Visual Basic.NET, your code will need to be upgraded before it can be used. When you try to open a Visual Basic 6 project in .NET, an Upgrade Wizard will automatically open and will step you through the upgrade process. It creates a new Visual Basic.NET project leaving the existing Visual Basic 6 project unchanged. If you have Visual Basic version 5 projects, it's best to upgrade them to 6 before moving on to 7 because the Upgrade Wizard assumes it is a version 6 project.

Upgrade Wizard

The first step in the conversion happens automatically when you open a Visual Basic 6.0 project in Visual Basic.NET. The Upgrade Wizard begins by converting the Visual Basic 6 forms to Windows Forms. This step is actually working quite well and without any noticeable difficulties. Next, the Upgrade Wizard makes changes to the syntax of the source code.

The code changes are the most difficult process that is involved in upgrading your projects. This is because certain objects and language features either have no equivalent in Visual Basic.NET, or have an equivalent too dissimilar for an automatic upgrade. Another problem with the upgrades is that they may not take advantage of the new features available in .NET.

Once the process is complete, Visual Basic.Net provides an "upgrade report" to help you make changes and review the status of your project. The items are displayed as tasks in the new Task List window, so you can easily see what changes are required and navigate to the code statement simply by double-clicking the task. Many times, the document recommendations simply represent good programming practices, but they also identify the Visual Basic 6 objects and methods that are no longer supported.

Automatic Code Upgrades

When your VB 6 code is upgraded to Visual Basic.NET, it follows a specific set of rules. The following rules list some basic information that you should keep in mind if you are currently planning to upgrade VB 6 project to .NET.

Variant to Object

Previous versions of Visual Basic supported the Variant data type, which could be assigned to any primitive type. In fact, it is the default data type if a variable wasn't declared in VB6. Visual Basic.NET converts variants to the Object data type.

Integer to Short

In Visual Basic.NET, the datatype for 16-bit whole numbers is now Short. The data type for 32-bit whole numbers is now Integer and Long is now 64 bits.

Here are a few examples:

VB6	VB.NET
Dim A as Integer	Dim A as Short
Dim B as Long	Dim B as Integer
N/A	Dim A as Long
Variant	N/A (Use new 'Object')
Currency	N/A (Use decimal or long)
N/A	Decimal
String	String (doesn't support fixed length strings)

APIs

The vast majority of API calls expect 32-bit values if they take numeric arguments. With the previous section in mind, you can see that problems are sure to arise. For instance, in VB6, a 32-bit value is a Long data type while in .NET, a Long is 64-bits. You'll have to use Integer as the data type in .NET to make the calls correctly. According to Microsoft documentation, many APIs will no longer be callable from VB or may have replacements.

The Upgrade Wizard tries to correct API calls by creating wrappers for them. This is not a good idea and you should look at every API call individually to make any changes you need.

Here is an example of an API call under each:

Visual Basic 6:
```
Private Declare Function GetVersion Lib "kernel32" ()
As Long
Function GetVer()
    Dim Ver As Long
    Ver = GetVersion()
    MsgBox ("System Version is " & Ver)
End Function
```

Visual Basic.NET:
```
Private Declare Function GetVersion Lib "kernel32" ()
As Integer
Function GetVer()
    Dim Ver As Integer
    Ver = GetVersion()
    MsgBox("System Version is " & Ver)
End Function
```

Newly Introduced Keywords

VB.NET introduces several new keywords that have no counterpart in VB6.

Keyword	Notes
Catch	New error handling – indicates code to use to process errors
Char	New character datatype
Finally	New error handling – indicates code to use to run regardless of errors
Imports	Makes an object hierarchy (namespace) available in a module
Inherits	Points to a base class for inheritance
MustOverride	Indicates that any class that derives from this class must supply an override for this member
MyBase	References the base class for use by subclass' code
Namespace	Specifies a namespace for a module
Overloads	Indicates there's more than one version of a function and the compiler can distinguish among them by the input parameters
Overrides	Indicates a member overrides the identically named member in the base class
Overridable	A member can be overridden in any class derived from the base class
Protected	This member is only available to classes derived from this class
ReadOnly	Used in a property that contains only a "Get"
Shared	All instances of a class should share a variable in a class
Throw	New error handling – used to raise an error
Try	New error handling – starts code with error handling enabled
Webmethod	Tags a method as part of a publicly available Web Service
WriteOnly	Used in a property that contains only a "Set"

Arrays

The use of arrays in VB.NET have also changed. In Visual Basic 6, if you declare like the following, you would get 11 items from 0 to 10:

Dim number(10) as Integer

With the VB.NET, the same array would only give you 10 items from 0 to 9.

Default Properties

In Visual Basic 6, a control or object had a default property that wouldn't need to be specified. For instance if you wanted to set a TextBox equal to a string, you would simply use:

txtInformation = "This is a string"

Visual Basic.NET doesn't support default properties. So instead, you have to make sure to specify it as follows:

txtInformation.Text = "This is a string"

References to Form Controls

Controls in Visual Basic 6 were public. That is, you could simply reference a control on form 1 inside the code window on form 2 by simply use form1.textbox1.text. In VB.NET, you'll have to create a public Let and Get property procedure for every control property you would like to have access to.

Get and Let are now combined in VB.NET, so instead of being two separate property procedures, you create one.

Visual Basic 6 Code:
```
Property Get PropertyA() As Integer
    m_PropertyA = PropertyA
End Property

Property Let PropertyA(NewValue As Integer)
    m_PropertyA = NewValue
End Property
```

Visual Basic.NET Code:
```
Property PropertyA() As Short
  Get
      m_PropertyA = MyPropertyA
  End Get
  Set
      m_PropertyA = Value
  End Set
End Property
```

Forms and Controls

Visual Basic.NET forms are now called Windows Forms. The following list will give you an idea of the differences between version 6 and .NET forms:

- Windows Forms do not support the OLE container control.
- Windows Forms only supports true-type and open-type fonts.
- There are no shape controls in Windows Forms. Shape controls have been upgraded to labels.
- Drap and drop properties of VB 6 do not work on Windows Forms.
- Windows Forms have no support for Dynamic Data Exchange (DDE).
- There is no line control in Windows Forms. Shape controls have been upgraded to labels.
- Windows Forms do not have access to form methods such as Circle or Line.
- Windows Forms do not support the PrintForm method.
- Clipboards are different and cannot be upgraded from version 6 to .NET.

Conclusion

This Appendix is a very basic overview of the changes and adjustments made in Visual Basic.NET. Microsoft has provided documentation that will aid you further in the upgrade process.

Appendix D

VIRTUAL-KEY CODES

This appendix will serve as a reference to the virtual keys that are often used in API functions. The following list does not include every key code but includes most of the common and the vast majority of less popular keys.

VK_LBUTTON
Left mouse button

VK_MBUTTON
Middle mouse button

VK_RBUTTON
Right mouse button

VK_BACK
Backspace

VK_TAB
Tab

VK_RETURN
Enter

VK_SHIFT
Shift keys

VK_CONTROL
Ctrl keys

VK_MENU
Alt keys

VK_PAUSE
Pause

VK_CANCEL
Cancel

VK_CAPITAL
Caps Lock

VK_ESCAPE
Esc

VK_SPACE
Spacebar

VK_PRIOR
Page Up

VK_NEXT
Page Down

VK_END
End

VK_HOME
Home

VK_LEFT
Left Arrow

VK_RIGHT
Right Arrow

VK_UP
Up Arrow

VK_DOWN
Down Arrow

VK_SELECT
Select

VK_SNAPSHOT
Print Screen

VK_INSERT
Insert

VK_DELETE
Delete

VK_HELP
Help

VK_0 0	**VK_H** H	**VK_Y** Y
VK_1 1	**VK_I** I	**VK_Z** Z
VK_2 2	**VK_J** J	**VK_STARTKEY** Start Menu
VK_3 3	**VK_K** K	**VK_CONTEXTKEY** Context Menu
VK_4 4	**VK_L** L	**VK_NUMPAD0** 0 on Number Pad
VK_5 5	**VK_M** M	**VK_NUMPAD1** 1 on Number Pad
VK_6 6	**VK_N** N	**VK_NUMPAD2** 2 on Number Pad
VK_7 7	**VK_O** O	**VK_NUMPAD3** 3 on Number Pad
VK_8 8	**VK_P** P	**VK_NUMPAD4** 4 on Number Pad
VK_9 9	**VK_Q** Q	**VK_NUMPAD5** 5 on Number Pad
VK_A A	**VK_R** R	**VK_NUMPAD6** 6 on Number Pad
VK_B B	**VK_S** S	**VK_NUMPAD7** 7 on Number Pad
VK_C C	**VK_T** T	**VK_NUMPAD8** 8 on Number Pad
VK_D D	**VK_U** U	**VK_NUMPAD9** 9 on Number Pad
VK_E E	**VK_V** V	**VK_MULTIPLY** *
VK_F F	**VK_W** W	**VK_ADD** +
VK_G G	**VK_X** X	**VK_DECIMAL** .

VK_DIVIDE
/

VK_F1
F1

VK_F2
F2

VK_F3
F3

VK_F4
F4

VK_F5
F5

VK_F6
F6

VK_F7
F7

VK_F8
F8

VK_F9
F9

VK_F10
F10

VK_F11
F11

VK_F12
F12

VK_F13
F13

VK_F14
F14

VK_F15
F15

VK_F16
F16

VK_F17
F17

VK_F18
F18

VK_F19
F19

VK_F20
F20

VK_F21
F21

VK_F22
F22

VK_F23
F23

VK_F24
F24

VK_NUMLOCK
Num Lock

VK_OEM_SCROLL
Scroll Lock

VK_OEM_1
;

VK_OEM_PLUS
=

VK_OEM_COMMA
,

VK_OEM_MINUS
-

VK_OEM_PERIOD
.

VK_OEM_2
/

VK_OEM_3
`

VK_OEM_4
[

VK_OEM_5
\

VK_OEM_6
]

VK_OEM_7
'

CONSTANT DEFINITIONS

These should be in the form of Const Definition when they are used (i.e. Const VK_LBUTTON=&H1)

VK_LBUTTON = &H1
VK_RBUTTON = &H2
VK_CANCEL = &H3
VK_MBUTTON = &H4
VK_BACK = &H8
VK_TAB = &H9
VK_CLEAR = &HC
VK_RETURN = &HD
VK_SHIFT = &H10
VK_CONTROL = &H11
VK_MENU = &H12
VK_PAUSE = &H13
VK_CAPITAL = &H14
VK_ESCAPE = &H1B
VK_SPACE = &H20
VK_PRIOR = &H21
VK_NEXT = &H22
VK_END = &H23
VK_HOME = &H24
VK_LEFT = &H25
VK_UP = &H26
VK_RIGHT = &H27
VK_DOWN = &H28
VK_SELECT = &H29
VK_PRINT = &H2A
VK_EXECUTE = &H2B
VK_SNAPSHOT = &H2C
VK_INSERT = &H2D
VK_DELETE = &H2E
VK_HELP = &H2F
VK_0 = &H30
VK_1 = &H31
VK_2 = &H32
VK_3 = &H33
VK_4 = &H34
VK_5 = &H35
VK_6 = &H36
VK_7 = &H37
VK_8 = &H38
VK_9 = &H39
VK_A = &H41
VK_B = &H42
VK_C = &H43
VK_D = &H44
VK_E = &H45
VK_F = &H46
VK_G = &H47
VK_H = &H48
VK_I = &H49
VK_J = &H4A
VK_K = &H4B
VK_L = &H4C
VK_M = &H4D
VK_N = &H4E
VK_O = &H4F
VK_P = &H50
VK_Q = &H51
VK_R = &H52
VK_S = &H53
VK_T = &H54
VK_U = &H55
VK_V = &H56
VK_W = &H57
VK_X = &H58
VK_Y = &H59
VK_Z = &H5A
VK_STARTKEY = &H5B
VK_CONTEXTKEY = &H5D
VK_NUMPAD0 = &H60
VK_NUMPAD1 = &H61
VK_NUMPAD2 = &H62
VK_NUMPAD3 = &H63
VK_NUMPAD4 = &H64
VK_NUMPAD5 = &H65
VK_NUMPAD6 = &H66
VK_NUMPAD7 = &H67
VK_NUMPAD8 = &H68
VK_NUMPAD9 = &H69
VK_MULTIPLY = &H6A
VK_ADD = &H6B
VK_SUBTRACT = &H6D
VK_DECIMAL = &H6E
VK_DIVIDE = &H6F
VK_F24 = &H87
VK_NUMLOCK = &H90
VK_OEM_SCROLL = &H91
VK_OEM_1 = &HBA
VK_OEM_PLUS = &HBB
VK_OEM_COMMA = &HBC
VK_OEM_MINUS = &HBD
VK_OEM_PERIOD = &HBE
VK_F1 = &H70
VK_F2 = &H71
VK_F3 = &H72
VK_F4 = &H73
VK_F5 = &H74

VK_F6 = &H75	VK_F14 = &H7D	VK_F22 = &H85
VK_F7 = &H76	VK_F15 = &H7E	VK_F23 = &H86
VK_F8 = &H77	VK_F16 = &H7F	VK_OEM_2 = &HBF
VK_F9 = &H78	VK_F17 = &H80	VK_OEM_3 = &HC0
VK_F10 = &H79	VK_F18 = &H81	VK_OEM_4 = &HDB
VK_F11 = &H7A	VK_F19 = &H82	VK_OEM_5 = &HDC
VK_F12 = &H7B	VK_F20 = &H83	VK_OEM_6 = &HDD
VK_F13 = &H7C	VK_F21 = &H84	VK_OEM_7 = &HDE

Appendix E

ASCII CHART

The use of ASCII characters is important in many VB projects. You can use the following chart if you need to look one of them up:

Value	Chr	Value	Chr	Value	Chr
32	Space	55	7	78	N
33	!	56	8	79	O
34	"	57	9	80	P
35	#	58	:	81	Q
36	$	59	;	82	R
37	%	60	<	83	S
38	&	61	=	84	T
39	'	62	>	85	U
40	(	63	?	86	V
41	)	64	@	87	W
42	*	65	A	88	X
43	+	66	B	89	Y
44	,	67	C	90	Z
45	-	68	D	91	[
46	.	69	E	92	\
47	/	70	F	93	]
48	0	71	G	94	^
49	1	72	H	95	_
50	2	73	I	96	`
51	3	74	J	97	a
52	4	75	K	98	b
53	5	76	L	99	c
54	6	77	M	100	d

Value	Chr	Value	Chr	Value	Chr
101	e	135	‡	169	©
102	f	136	ˆ	170	ª
103	g	137	‰	171	«
104	h	138	Š	172	¬
105	i	139	‹	173	-
106	j	140	Œ	174	®
107	k	141		175	¯
108	l	142	Ž	176	°
109	m	143		177	±
110	n	144		178	²
111	o	145	'	179	³
112	p	146	'	180	´
113	q	147	"	181	µ
114	r	148	"	182	¶
115	s	149	•	183	·
116	t	150	–	184	¸
117	u	151	—	185	¹
118	v	152	˜	186	º
119	w	153	™	187	»
120	x	154	š	188	¼
121	y	155	›	189	½
122	z	156	œ	190	¾
123	{	157		191	¿
124	|	158	ž	192	À
125	}	159	Ÿ	193	Á
126	~	160		194	Â
127	▓	161	¡	195	Ã
128	€	162	¢	196	Ä
129		163	£	197	Å
130	‚	164	¤	198	Æ
131	ƒ	165	¥	199	Ç
132	„	166	¦	200	È
133	…	167	§	201	É
134	†	168	¨	202	Ê

Value	Chr	Value	Chr	Value	Chr
203	Ë	223	Ý	243	ó
204	Ì	224	Þ	244	ô
205	Í	225	á	245	õ
206	Î	226	â	246	ö
207	Ï	227	ã	247	÷
208	Ð	228	ä	248	ø
209	Ñ	229	å	249	ù
210	Ò	230	æ	250	ú
211	Ó	231	ç	251	û
212	Ô	232	è	252	ü
213	Õ	233	é	253	ý
214	Ö	234	ê	254	þ
215	×	235	ë	255	ÿ
216	Ø	236	ì		
217	Ù	237	í		
218	Ú	238	î		
219	Û	239	ï		
220	Ü	240	ð		
221		241	ñ		
222		242	ò		

Appendix F

VISUAL BASIC FUNCTIONS AND STATEMENTS

This chapter serves as a list of some of the common and not so common built-in functions and statements that Visual Basic offers. For additional information, you should consult the online Help in Visual Basic or the Microsoft Language Reference.

VISUAL BASIC FUNCTIONS

Abs

`Abs(number)`

Returns a positive value of a number or formula, regardless of its actual positive or negative value. The data type returned is the same as the data type of the number argument.

Array

`Array(arglist)`

Returns a Variant that contains an array. The arglist, which is required, refers to a comma-delimited list of values that make up the elements of the array, with the first value corresponding to the first element of the array, the second value corresponding to the second element of the array, etc.

Asc

`Asc(string)`

Returns an Integer represents the ASCII character code corresponding to the first letter in a string.

Atn

`Atn(number)`

Returns a Double that is the arctangent of number.

CBool

`CBool(expression)`

Converts the value of expression to Boolean. If the expression evaluates to a nonzero value, CBool returns True; otherwise, it returns False.

CByte

`CByte(expression)`

Converts the value of expression to Byte. The expression must be a numeric value between 0 and 255.

CCur

`CCur(expression)`

Converts the value of expression to Currency. The expression must be a numeric value between -922,337,203,685,477.5808 and 922,337,203,685,477.5807.

CDate

`CDate(expression)`

Converts the value of expression to Date. The expression must be a valid date expression.

CDbl

`CDbl(expression)`

Converts the value of expression to Double. The expression must be a numeric value between -1.79769313486232E308 and -4.94065645841247E-324 for expressions less than 0, or between 4.94065645841247E-324 and 1.79769313486232E308 for positive.

CDec

`CDec(expression)`

Converts the value of expression to Decimal. The expression must be a numeric value of +/-79,228,162,514,264,337,593,543,950,335 for numbers without decimal places or +/-7.9228162514264337593543950335 for others. The smallest possible non-zero number is 0.0000000000000000000000000001.

CInt

`CInt(expression)`

Converts the value of expression to Integer. The expression must be numeric value between -32,768 to 32,768. All fractions are rounded.

CLng

`CLng(expression)`

Converts the value of expression to Long. The expression must be numeric value between -2,147,483,648 to 2,147,483,647. All fractions are rounded.

CSng

`CSng(expression)`

Converts the value of expression to Single. The expression must be numeric value between -3.402823E38 to -1.401298E-45 for negative values or 1.401298E-45 to 3.402823E38 for positive values. All fractions are rounded.

CVar

`CSng(expression)`

Converts the value of expression to Variant. Same range as Double for numerics or same range as String for non-numerics.

Chr

`Chr(charcode)`

Returns a one-character String value that represents the ASCII character of charcode.

Command

`Command`

Returns any command-line arguments specified when launching Visual Basic or if a compiled program it returns the command-line arguments specified when the program was launched.

Cos

`Cos(number)`

Returns a Double value that is the cosine of the number.

CreateObject

`CreateObject(class)`

Creates and returns a reference to an ActiveX object. If an object has registered itself as a single-instance object, only one instance of the object is created regardless of the number of attempts.

CurDir

`CurDir[(drive)]`

Returns a String that represents the path of the current directory.

Date

`Date`

Returns a Variant that contains the current system date.

DateAdd

`DateAdd(interval, number, date)`

Returns a Variant that is calculated by taking the date specified by the date argument and adding or subtracting the amount of time.

DateDiff

`DateDiff(interval, date1, date2[,firstdayofweek[, firstweekofyear]])`

Returns a Variant that represents the number of time units between two dates (date1 and date2).

DatePart

`DatePart(interval, date[,firstdayofweek[, firstweekofyear]])`

Returns a Variant that contains the part of date specified by interval.

DateSerial

`DateSerial(year, month, day)`

Returns a Variant that represents a date as specified by the year, month, and day arguments.

DateValue

`DateValue(date)`

Returns a Variant that is derived from the date value specified by the date argument.

Day

`Day(date)`

Returns a Variant that represents the day of the month for the date value specified by the date argument.

DDB

`DDB(cost, salvage, life, period[, factor])`

Returns a Double that represents the depreciation of an asset for a specified amount of time.

Dir

`Dir[(pathname[, attributes])]`

Returns a String that contains the name of a file, directory, or folder that matches the pattern specified in the pathname argument. An optional file attribute can also be matched.

DoEvents

`DoEvents( )`

Gives control to Windows so that it can process other events.

EOF

`EOF(filenumber)`

Returns an Integer (1 or 0) that indicates whether the end of file marker has been reached for the Random or Input file.

Error

`Error[(errornumber)]`

Returns a String that contains the message associated with errornumber.

FileAttr

`FileAttr(filenumber, returntype)`

Returns a Long value that indicates the file mode for a file opened using the Open.

FileDateTime

`FileDateTime(pathname)`

Returns a Variant that indicates the date and time when a file was last modified.

FileLen

`FileLen(pathname)`

Returns a Long that contains the file size.

Fix

`Fix(number)`

Returns the Integer portion of the number.

FV

`FV(rate, nper, pmt[, pv[, type]])`

Returns a Double that indicates the future value of an annuity based on a number (nper) of periodic fixed payment amounts (pmt) and a fixed interest rate (rate).

GetAllSettings

`GetAllSettings(appname, section)`

Returns a list of key settings and their values from a specific application entry and section in the System Registry.

GetAttr

`GetAttr(pathname)`

Returns an Integer that represents the attributes for the file, directory, or folder.

GetObject

`GetObject([pathname] [,class])`

Returns a reference to an ActiveX object from a file.

Hex

`Hex(number)`

Returns a String value that represents the hexadecimal value.

Hour

`Hour(time)`

Returns a Variant that represents the hour between 0 and 23.

IIf

`IIf(expression, truepart, falsepart)`

Returns one of two values based on whether expression evaluates to True or False. If it is True, then returns the truepart, or if it is False, then the falsepart is returned.

Input

`Input(number, [#]filenumber)`

Returns a String value containing characters read in from an open file.

InputBox

`InputBox(prompt[, title][, default][, xpos][, _ ypos][,helpfile, context])`

Displays a dialog box that waits for the user to enter text or click a button. It returns what the user entered as a String value.

InStr

`InStr([start, ]string1, string2[, compare])`

Returns a Variant that specifies the starting position of the first occurrence of a string (string2) within another string (string1).

Int

`Int(number)`

Returns the integer portion of the number.

IPmt

`IPmt(rate, per, nper, pv[, fv[, type]])`

Returns a Double that indicates the interest payment for a fixed-period annuity based on a number (nper) of periodic fixed payments (per) and a fixed interest rate (rate).

IRR

`IRR(values()[, guess])`

Returns a Double indicating the internal rate of return for an array of values that represent cash flow.

IsArray

`IsArray(varname)`

Returns a Boolean that indicates whether the variable is an array.

IsDate

`IsDate(expression)`

Returns a Boolean that indicates whether expression is capable of being converted to a date value.

IsEmpty

`IsEmpty(expression)`

Returns a Boolean that indicates whether a numeric or string expression has been initialized.

IsError

`IsError(expression)`

Returns a Boolean that indicates whether a given expression is an error value.

IsMissing

`IsMissing(argname)`

Returns a Boolean that indicates whether an optional Variant argument (argname) has been passed to a procedure.

IsNull

`IsNull(expression)`

Returns a Boolean that indicates whether a given expression contains no data.

IsNumeric

`IsNumeric(expression)`

Returns a Boolean that indicates whether a given expression can be evaluated as a numeric value.

IsObject

`IsObject(identifier)`

Returns a Boolean that indicates whether a given identifier represents an object variable.

LBound

`LBound(arrayname[, dimension])`

Returns a Long that represents the smallest subscript for a dimensioned array.

LCase

`LCase(string)`

Converts String to all lowercase and returns the resulting String.

Left

`Left(string, length)`

Returns a String of a certain length that is taken from the left side of a given string.

Len

Len(string | varname)

Returns a Long that indicates the number of characters in a string.

LoadPicture

LoadPicture([stringexpression])

Loads the image specified by the stringexpression argument and returns it.

LoadResData

LoadResData(index, format)

Loads data from a resource (.RES) file.

LoadResPicture

LoadResPicture(index, format)

Loads a bitmap, icon, or cursor from the resource (.RES) file.

LoadResString

LoadResString(index)

Loads a string from the resource (.RES) file.

Loc

Loc(filenumber)

Returns a Long that indicates the current byte position within an open file.

LOF

LOF(filenumber)

Returns a Boolean that represents the byte size of an open file.

Log

Log(number)

Returns a Double that represents the natural log.

LTrim

LTrim(string)

Returns a Variant that contains a copy of a given string with any leading spaces removed.

Mid

Mid(string, start[, length])

Returns a String of one or more characters, taken from a String beginning at a certain position and a given length.

Minute

Minute(time)

Returns a Variant that represents the minute (0-59) of the time.

Month

Month(date)

Returns a Variant that represents the month.

MsgBox

MsgBox(prompt[, buttons][, title][, helpfile, context]

Displays a message in a dialog box with one or more buttons and waits for the user to respond.

Now

`Now`

Returns a Variant that contains the current system date and time.

Oct

`Oct(number)`

Returns a String that represents the octal value.

RaiseEvent

`RaiseEvent eventname [(argumentlist)]`

Triggers an event.

Rate

`Rate(nper, pmt, pv[, fv[, type[, guess]]])`

Returns a Double value that indicates the fixed interest rate per period for an annuity based on a number (nper) of periodic fixed payments (pmt).

RGB

`RGB(red, green, blue)`

Returns a Long that represents an RGB color value.

Right

`Right(string, length)`

Returns a String of characters taken from the right side of a given string.

Rnd

`Rnd[(number)]`

Returns a Single that contains a randomly generated number less than 1 but greater than or equal to zero.

RTrim

`RTrim(string)`

Returns a Variant that contains a copy of a given string with any trailing spaces removed.

Second

`Second(time)`

Returns a Variant that represents the second of a given time.

Seek

`Seek(filenumber)`

Returns a Long that specifies the current record or byte position for an open file.

Sgn

`Sgn(number)`

Returns a Variant that represents the sign of a given number.

Shell

`Shell(pathname[, windowstyle])`

Runs the executable program specified by the pathname argument and returns a Variant that represents the program's task ID.

Sin

`Sin(number)`

Returns a Double that represents the sine of a given angle.

SLN

`SLN(cost, salvage, life)`

Returns a Double that represents the straight-line depreciation of an asset.

Space

`Space(number)`

Returns a Variant that contains a number of spaces.

Spc

`Spc(n)`

Inserts a specified number of spaces when writing or displaying text using the Print # or the Print method.

Sqr

`Sqr(number)`

Returns a Double that represents the square root.

Str

`Str(number)`

Returns a Variant that is a representation of a given number.

StrComp

`StrComp(string1, string2[, compare])`

Returns a Variant that indicates the result of a comparison between two strings (string1 and string2).

StrConv

`StrConv(string, conversion)`

Returns a Variant that has been converted from an original string as specified by the conversion argument.

String

`String(number, character)`

Returns a Variant of given length that is filled with a given character.

Tab

`Tab(n)`

Positions output to a given column when writing or displaying text using Print # or the Print method.

Tan

`Tan(number)`

Returns a Double that represents the tangent.

Time

`Time`

Returns a Variant that contains the current system time.

Timer

`Timer`

Returns a Single that represents the number of seconds that have elapsed since midnight.

TimeSerial

`TimeSerial(hour, minute, second)`

Returns a Variant that represents a time as specified by the hour, minute, and second arguments.

TimeValue

`TimeValue(time)`

Returns a Variant that is derived from the time value.

Trim

`Trim(string)`

Returns a Variant that contains a copy of a given string with any leading and trailing spaces removed.

TypeName

`TypeName(varname)`

Returns a String that indicates the data type of a given variable.

UBound

`UBound(arrayname[, dimension])`

Returns a Long that represents the largest subscript for a dimensioned array.

UCase

`UCase(string)`

Converts a string to all uppercase and returns the resulting String.

Val

`Val(string)`

Returns the numeric value of a string.

VarType

`VarType(varname)`

Returns an Integer that represents the subtype of a variable.

Weekday

`Weekday(date, [firstdayofweek])`

Returns a Variant that represents the day of the week for a given date.

Year

`Year(date)`

Returns a Variant that represents the year.

VISUAL BASIC STATEMENTS

AppActivate
`AppActivate title[,wait]`

Activates the application window that has the string title in its title bar.

Beep
`Beep`

Sounds a tone through the PC's speaker.

Call
`[Call] name [argumentlist]`

Executes a sub, function, or DLL procedure. The Call keyword is optional, but if it is included, then at least one or more arguments for argumentlist must also be included.

ChDir
`ChDir path`

Changes the current directory.

ChDrive
`ChDrive drive`

Changes the current drive.

Close
`Close [filenumberlist]`

Closes any files opened with the Open.

Const
`[Public | Private] Const constname [As type] = expression`

Declares a constant..

Date
`Date = date`

Sets the current system.

Declare
`[Public | Private] Declare Function name Lib "libname" [Alias"aliasname"][([arglist])] [As type]`

Declares references to Sub or Function procedures in an external DLL (dynamic-link library).

DeleteSetting
`DeleteSetting appname, section[, key]`

Deletes an application's section or key setting entries from the System Registry.

Dim
`Dim varname As vartype`

Declares one or more variables or objects.

Do...Loop
`Do [{While | Until} condition]`

Repeats one or more statements while a condition is True or until a condition becomes True.

End

`End xyz`

Ends a program, procedure, type structure, or program block specified in xyz, i.e. End, End Function, End If, End Property, End Select, End Sub, End Type, etc.

Enum

`[Public | Private] Enum name`

Declares an enumeration type named name that is composed of one or more members.

Erase

`Erase arraylist`

Reinitializes the elements in an array and frees up the dynamic-array storage space that was taken up by the array.

Error

`Error errornumber`

Causes an error to occur.

Event

`[Public] Event procedurename [(arglist)]`

Declares a user-defined event with the name procedurename.

Exit

`Exit xyz`

Exits a procedure or looping structure specificed in xyz.

FileCopy

`FileCopy source, destination`

Copies a file from source to destination.

For...Next

`For counter = start To end [Step step]`

Executes one or more statements a specified number of times in a given amount of steps.

Function

`Function xyz`

Declares the various parts of a Function procedure.

Get

`Get [#]filenumber,[ recnumber,] varname`

Reads data from the open disk file.

If...Then...Else

`If condition Then [ s] [Else else s]`

Conditionally executes one or more statements if the value expressed by condition is True.

Implements

`Implements [interfacename | class]`

Specifies an interface or class that will be implemented in a class module.

Input

`Input #filenumber, varlist`

Reads data from an open sequential file and assigns the data to variables.

Kill

`Kill pathname`

Deletes the file or directory represented by the pathname argument.

Let

`[Let] varname = expression`

Assigns the value of an expression to a variable. Let is assumed and therefor is not used regularly.

Line Input

`Line Input #filenumber, varname`

Reads a line of data from an open disk file that is placed in variables.

Load

`Load object`

Loads an object, such as a control, into memory.

LSet

`LSet stringvar = string`

Assigns the value of an expression to a variable or property.

Mid

`Mid(stringvar, start[, length]) = string`

Replaces one or more characters in a String variable with another string.

MkDir

`MkDir path`

Creates the new directory.

Name

`Name oldpathname As newpathname`

Renames a file, directory, or folder.

Open

`Open pathname For mode [Access access] [lock] As [#]filenumber [Len=reclength]`

Enables input/output (I/O) to a file.

Option Base

`Option Base [0 | 1]`

Used at module level to declare the default lower bound for array subscripts.

Option Compare

`Option Compare [Binary | Text | Database]`

Used at module level to declare the default comparison method to use when string data is compared.

Option Explicit

`Option Explicit`

Forces explicit declaration of all variables in a module.

Option Private

`Option Private Module`

Prevents a module's contents from being used outside its project.

Print

`Print #filenumber, [outputlist]`

Writes data to the open sequential file.

Private

`Private [WithEvents] varname[([subscripts])] [As [New] type] [,[WithEvents] varname[([subscripts])] [As [New] type]] . . .`

Used at module level to declare private variables and allocate storage space.

Property Get

`[Public | Private | Friend] [Static] Property Get name [(arglist)] [As type]`

Declares name, arguments, and code that form the body of a Property procedure.

Public

`Public [WithEvents] varname[([subscripts])] [As [New] type][,[WithEvents] varname[([subscripts])] [As [New] type]]...`

Declares one or more public variables.

Put

`Put [#]filenumber, [recnumber], varname`

Writes data to the open disk file.

Randomize

`Randomize [number]`

Initializes the random number generator, using the optional number argument as a seed value.

ReDim

`ReDim [Preserve] varname(subscripts) [As type] [, varname(subscripts) [As type]] . . .`

Reallocate storage space for dynamic array variables.

Rem

`Rem comments`

Allows comments to be added to a program. Can also use an apostrophe.

Reset

`Reset`

Closes all files opened with the Open and writes any file buffer contents to disk.

Resume

`Resume xyz`

Resumes execution of a program when an error-handling routine is finished.

RmDir
`RmDir path`

Removes a folder or directory.

RSet
`RSet stringvar = string`

Assigns a string value to a String variable and right-aligns the string to the variable.

SavePicture
`SavePicture picture, stringexpression`

Saves graphic image from an object's Picture or Image property to a file.

SaveSetting
`SaveSetting appname, section, key, setting`

Saves an application name, section, key setting and value in the System Registry.

Seek
`Seek [#]filenumber, position`

Sets the record or byte position of an open file.

Select Case
`Select Case xyz`

Evaluates an expression and depending on the result, executes one or more statements that correspond to the result.

SendKeys
`SendKeys xyz`

Generates one or more keystrokes as if they came from the keyboard.

Set
`Set objectvar = {[New] objectexpression | Nothing}`

Assigns an object reference to a variable or property.

SetAttr
`SetAttr pathname, attributes`

Sets attributes for a file or directory.

Static
`Static varname[([subscripts])] [As [New] type] [, varname[([subscripts])] [As [New] type] ...`

Declares one or more static variables.

Stop
`Stop`

Suspends program execution.

Sub
`[Public | Private | Friend] [Static] Sub name [(arglist)]`

Declares the various parts of a Sub procedure.

Type
`[Private | Public] Type xyz`

Defines a user-defined type (UDT) structure that contains one or more elements.

Unload

```
Unload object
```

Unloads an object from memory and frees up any resources being used by the object.

Unlock

```
Unlock [#]filenumber[,
recordrange]
```

Removes locking of an open file.

While...Wend

```
While condition
    [ s]
Wend
```

Repeats one or more statements while a condition remains True.

Width

```
Width #filenumber, width
```

Assigns output line width for an open file.

With

```
With object
    [ s]
End With
```

Executes one or more statements on a single object or user-defined type.

Write

```
Write #filenumber,
[outputlist]
```

Writes data to the open sequential file.

Appendix G

API CONSTANTS A THROUGH M

ABORTDOCAPI
Global Const ABORTDOCAPI = 2

ABSOLUTE
Global Const ABSOLUTE = 1

ALTERNATE
Global Const ALTERNATE = 1

ANSI_CHARSET
Global Const ANSI_CHARSET = 0

ANSI_FIXED_FONT
Global Const ANSI_FIXED_FONT = 11

ANSI_VAR_FONT
Global Const ANSI_VAR_FONT = 12

ASPECT_FILTERING
Global Const ASPECT_FILTERING = &H1

ASPECTX
Global Const ASPECTX = 40

ASPECTXY
Global Const ASPECTXY = 44

ASPECTY
Global Const ASPECTY = 42

ANDINFO
Global Const BANDINFO = 24

BEGIN_PATH
Global Const BEGIN_PATH = 4096

BI_RGB
Global Const BI_RGB = 0&

BI_RLE4
Global Const BI_RLE4 = 2&

BI_RLE8
Global Const BI_RLE8 = 1&

BITSPIXEL
Global Const BITSPIXEL = 12

BLACK_BRUSH
Global Const BLACK_BRUSH = 4

BLACK_PEN
Global Const BLACK_PEN = 7

BLACKNESS
Global Const BLACKNESS = &H42&

BLACKONWHITE
Global Const BLACKONWHITE = 1

BM_GETCHECK
Global Const BM_GETCHECK = WM_USER+0

BM_GETSTATE
Global Const BM_GETSTATE = WM_USER+2

BM_SETCHECK
Global Const BM_SETCHECK = WM_USER+1

BM_SETSTATE
Global Const BM_SETSTATE = WM_USER+3

BM_SETSTYLE
Global Const BM_SETSTYLE = WM_USER+4

BN_CLICKED
Global Const BN_CLICKED = 0

BN_DISABLE
Global Const BN_DISABLE = 4

BN_DOUBLECLICKED
Global Const BN_DOUBLECLICKED = 5

BN_HILITE
Global Const BN_HILITE = 2

BN_PAINT
Global Const BN_PAINT = 1

BN_UNHILITE
Global Const BN_UNHILITE = 3

BS_3STATE
Global Const BS_3STATE = &H5&

BS_AUTO3STATE
Global Const BS_AUTO3STATE = &H6&

BS_AUTOCHECKBOX
Global Const BS_AUTOCHECKBOX = &H3&

BS_AUTORADIOBUTTON
Global Const BS_AUTORADIOBUTTON = &H9&

BS_CHECKBOX
Global Const BS_CHECKBOX = &H2&

BS_DEFPUSHBUTTON
Global Const BS_DEFPUSHBUTTON = &H1&

BS_DIBPATTERN
Global Const BS_DIBPATTERN = 5

BS_GROUPBOX
Global Const BS_GROUPBOX = &H7&

BS_HATCHED
Global Const BS_HATCHED = 2

BS_HOLLOW
Global Const BS_HOLLOW = BS_NULL

BS_INDEXED
Global Const BS_INDEXED = 4

BS_LEFTTEXT
Global Const BS_LEFTTEXT = &H20&

BS_NULL
Global Const BS_NULL = 1

BS_OWNERDRAW
Global Const BS_OWNERDRAW = &HB&

BS_PATTERN
Global Const BS_PATTERN = 3

BS_PUSHBOX
Global Const BS_PUSHBOX = &HA&

BS_PUSHBUTTON
Global Const BS_PUSHBUTTON = &H0&

BS_RADIOBUTTON
Global Const BS_RADIOBUTTON = &H4&

BS_SOLID
Global Const BS_SOLID = 0

BS_USERBUTTON
Global Const BS_USERBUTTON = &H8&

B_ADDSTRING
Global Const CB_ADDSTRING = (WM_USER+3)

CB_DELETESTRING
Global Const CB_DELETESTRING = (WM_USER+4)

CB_DIR
Global Const CB_DIR = (WM_USER+5)

CB_ERR
Global Const CB_ERR = (-1)

CB_ERRSPACE
Global Const CB_ERRSPACE = (-2)

CB_FINDSTRING
Global Const CB_FINDSTRING = (WM_USER+12)

CB_FINDSTRINGEXACT
Global Const CB_FINDSTRINGEXACT = (WM_USER+24)

CB_GETCOUNT
Global Const CB_GETCOUNT = (WM_USER+6)

CB_GETCURSEL
Global Const CB_GETCURSEL = (WM_USER+7)

CB_GETDROPPEDCONTROLRECT
Global Const CB_GETDROPPEDCONTROLRECT = (WM_USER+18)

CB_GETDROPPEDSTATE
Global Const CB_GETDROPPEDSTATE = (WM_USER+23)

CB_GETEDITSEL
Global Const CB_GETEDITSEL = (WM_USER+0)

CB_GETEXTENDEDUI
Global Const CB_GETEXTENDEDUI = (WM_USER+22)

CB_GETITEMDATA
Global Const CB_GETITEMDATA = (WM_USER+16)

CB_GETITEMHEIGHT
Global Const CB_GETITEMHEIGHT = (WM_USER+20)

CB_GETLBTEXT
Global Const CB_GETLBTEXT = (WM_USER+8)

CB_GETLBTEXTLEN
Global Const CB_GETLBTEXTLEN = (WM_USER+9)

CB_INSERTSTRING
Global Const CB_INSERTSTRING = (WM_USER+10)

CB_LIMITTEXT
Global Const CB_LIMITTEXT = (WM_USER+1)

CB_MSGMAX
Global Const CB_MSGMAX = (WM_USER+19)

CB_OKAY
Global Const CB_OKAY = 0

CB_RESETCONTENT
Global Const CB_RESETCONTENT = (WM_USER+11)

CB_SELECTSTRING
Global Const CB_SELECTSTRING = (WM_USER+13)

CB_SETCURSEL
Global Const CB_SETCURSEL = (WM_USER+14)

CB_SETEDITSEL
Global Const CB_SETEDITSEL = (WM_USER+2)

CB_SETEXTENDEDUI
Global Const CB_SETEXTENDEDUI = (WM_USER+21)

CB_SETITEMDATA
Global Const CB_SETITEMDATA = (WM_USER+17)

CB_SETITEMHEIGHT
Global Const CB_SETITEMHEIGHT = (WM_USER+19)

CB_SHOWDROPDOWN
Global Const CB_SHOWDROPDOWN = (WM_USER+15)

CBM_INIT
Global Const CBM_INIT = &H4&

CBN_CLOSEUP
Global Const CBN_CLOSEUP = 8

CBN_DBLCLK
Global Const CBN_DBLCLK = 2

CBN_DROPDOWN
Global Const CBN_DROPDOWN = 7

CBN_EDITCHANGE
Global Const CBN_EDITCHANGE = 5

CBN_EDITUPDATE
Global Const CBN_EDITUPDATE = 6

CBN_ERRSPACE
Global Const CBN_ERRSPACE = (-1)

CBN_KILLFOCUS
Global Const CBN_KILLFOCUS = 4

CBN_SELCHANGE
Global Const CBN_SELCHANGE = 1

CBN_SELENDCANCEL
Global Const CBN_SELENDCANCEL = 10

CBN_SELENDOK
Global Const CBN_SELENDOK = 9

CBN_SETFOCUS
Global Const CBN_SETFOCUS = 3

CBR_110
Global Const CBR_110 = &HFF10

CBR_1200
Global Const CBR_1200 = &HFF13

CBR_128000
Global Const CBR_128000 = &HFF23

CBR_14400
Global Const CBR_14400 = &HFF17

CBR_19200
Global Const CBR_19200 = &HFF18

CBR_2400
Global Const CBR_2400 = &HFF14

CBR_256000
Global Const CBR_256000 = &HFF27

CBR_300
Global Const CBR_300 = &HFF11

CBR_38400
Global Const CBR_38400 = &HFF1B

CBR_4800
Global Const CBR_4800 = &HFF15

CBR_56000
Global Const CBR_56000 = &HFF1F

CBR_600
Global Const CBR_600 = &HFF12

CBR_9600
Global Const CBR_9600 = &HFF16

CBS_AUTOHSCROLL
Global Const CBS_AUTOHSCROLL = &H40&

CBS_DISABLENOSCROLL
Global Const CBS_DISABLENOSCROLL = &H0800&

CBS_DROPDOWN
Global Const CBS_DROPDOWN = &H2&

CBS_DROPDOWNLIST
Global Const CBS_DROPDOWNLIST = &H3&

CBS_HASSTRINGS
Global Const CBS_HASSTRINGS = &H200&

CBS_NOINTEGRALHEIGHT
Global Const CBS_NOINTEGRALHEIGHT = &H400&

CBS_OEMCONVERT
Global Const CBS_OEMCONVERT = &H80&

CBS_OWNERDRAWFIXED
Global Const CBS_OWNERDRAWFIXED = &H10&

CBS_OWNERDRAWVARIABLE
Global Const CBS_OWNERDRAWVARIABLE = &H20&

CBS_SIMPLE
Global Const CBS_SIMPLE = &H1&

CBS_SORT
Global Const CBS_SORT = &H100&

CC_CHORD
Global Const CC_CHORD = 4

CC_CIRCLES
Global Const CC_CIRCLES = 1

CC_ELLIPSES
Global Const CC_ELLIPSES = 8

CC_INTERIORS
Global Const CC_INTERIORS = 128

CC_NONE
Global Const CC_NONE = 0

CC_PIE
Global Const CC_PIE = 2

CC_STYLED
Global Const CC_STYLED = 32

CC_WIDE
Global Const CC_WIDE = 16

CC_WIDESTYLED
Global Const CC_WIDESTYLED = 64

CE_BREAK
Global Const CE_BREAK = &H10

CE_CTSTO
Global Const CE_CTSTO = &H20

CE_DNS
Global Const CE_DNS = &H800

CE_DSRTO
Global Const CE_DSRTO = &H40

CE_FRAME
Global Const CE_FRAME = &H8

CE_IOE
Global Const CE_IOE = &H400

CE_MODE
Global Const CE_MODE = &H8000

CE_OOP
Global Const CE_OOP = &H1000

CE_OVERRUN
Global Const CE_OVERRUN = &H2

CE_PTO
Global Const CE_PTO = &H200

CE_RLSDTO
Global Const CE_RLSDTO = &H80

CE_RXOVER
Global Const CE_RXOVER = &H1

CE_RXPARITY
Global Const CE_RXPARITY = &H4

CE_TXFULL
Global Const CE_TXFULL = &H100

CF_BITMAP
Global Const CF_BITMAP = 2

CF_DIB
Global Const CF_DIB = 8

CF_DIF
Global Const CF_DIF = 5

CF_DSPBITMAP
Global Const CF_DSPBITMAP = &H82

CF_DSPMETAFILEPICT
Global Const CF_DSPMETAFILEPICT = &H83

CF_DSPTEXT
Global Const CF_DSPTEXT = &H81

CF_GDIOBJFIRST
Global Const CF_GDIOBJFIRST = &H300

CF_GDIOBJLAST
Global Const CF_GDIOBJLAST = &H3FF

CF_METAFILEPICT
Global Const CF_METAFILEPICT = 3

CF_OEMTEXT
Global Const CF_OEMTEXT = 7

CF_OWNERDISPLAY
Global Const CF_OWNERDISPLAY = &H80

CF_PALETTE
Global Const CF_PALETTE = 9

CF_PRIVATEFIRST
Global Const CF_PRIVATEFIRST = &H200

CF_PRIVATELAST
Global Const CF_PRIVATELAST = &H2FF

CF_SYLK
Global Const CF_SYLK = 4

CF_TEXT
Global Const CF_TEXT = 1

CF_TIFF
Global Const CF_TIFF = 6

CLIP_CHARACTER_PRECIS
Global Const CLIP_CHARACTER_PRECIS = 1

CLIP_DEFAULT_PRECIS
Global Const CLIP_DEFAULT_PRECIS = 0

CLIP_EMBEDDED
Global Const CLIP_EMBEDDED = &H80

CLIP_LH_ANGLES
Global Const CLIP_LH_ANGLES = &H10

CLIP_STROKE_PRECIS
Global Const CLIP_STROKE_PRECIS = 2

CLIP_TO_PATH
Global Const CLIP_TO_PATH = 4097

CLIP_TT_ALWAYS
Global Const CLIP_TT_ALWAYS = &H20

CLIPCAPS
Global Const CLIPCAPS = 36

CLRDTR
Global Const CLRDTR = 6

CLRRTS
Global Const CLRRTS = 4

CN_EVENT
Global Const CN_EVENT = &H0004

CN_RECEIVE
Global Const CN_RECEIVE = &H0001

CN_TRANSMIT
Global Const CN_TRANSMIT = &H0002

COLOR_ACTIVEBORDER
Global Const COLOR_ACTIVEBORDER = 10

COLOR_ACTIVECAPTION
Global Const COLOR_ACTIVECAPTION = 2

COLOR_APPWORKSPACE
Global Const COLOR_APPWORKSPACE = 12

COLOR_BACKGROUND
Global Const COLOR_BACKGROUND = 1

COLOR_BTNFACE
Global Const COLOR_BTNFACE = 15

COLOR_BTNHIGHLIGHT
Global Const COLOR_BTNHIGHLIGHT = 20

COLOR_BTNSHADOW
Global Const COLOR_BTNSHADOW = 16

COLOR_BTNTEXT
Global Const COLOR_BTNTEXT = 18

COLOR_CAPTIONTEXT
Global Const COLOR_CAPTIONTEXT = 9

COLOR_ENDCOLORS
Global Const COLOR_ENDCOLORS = COLOR_BTNTEXT

COLOR_GRAYTEXT
Global Const COLOR_GRAYTEXT = 17

COLOR_HIGHLIGHT
Global Const COLOR_HIGHLIGHT = 13

COLOR_HIGHLIGHTTEXT
Global Const COLOR_HIGHLIGHTTEXT = 14

COLOR_INACTIVEBORDER
Global Const COLOR_INACTIVEBORDER = 11

COLOR_INACTIVECAPTION
Global Const COLOR_INACTIVECAPTION = 3

COLOR_INACTIVECAPTIONTEXT
Global Const COLOR_INACTIVECAPTIONTEXT = 19

COLOR_MENU
Global Const COLOR_MENU = 4

COLOR_MENUTEXT
Global Const COLOR_MENUTEXT = 7

COLOR_SCROLLBAR
Global Const COLOR_SCROLLBAR = 0

COLOR_WINDOW
Global Const COLOR_WINDOW = 5

COLOR_WINDOWFRAME
Global Const COLOR_WINDOWFRAME = 6

COLOR_WINDOWTEXT
Global Const COLOR_WINDOWTEXT = 8

COLORONCOLOR
Global Const COLORONCOLOR = 3

COLORRES
Global Const COLORRES = 108

COMPLEXREGION
Global Const COMPLEXREGION = 3

CP_NONE
Global Const CP_NONE = 0

CP_RECTANGLE
Global Const CP_RECTANGLE = 1

CS_BYTEALIGNCLIENT
Global Const CS_BYTEALIGNCLIENT = &H1000

CS_BYTEALIGNWINDOW
Global Const CS_BYTEALIGNWINDOW = &H2000

CS_CLASSDC
Global Const CS_CLASSDC = &H40

CS_DBLCLKS
Global Const CS_DBLCLKS = &H8

CS_GLOBALCLASS
Global Const CS_GLOBALCLASS = &H4000

CS_HREDRAW
Global Const CS_HREDRAW = &H2

CS_KEYCVTWINDOW
Global Const CS_KEYCVTWINDOW = &H4

CS_NOCLOSE
Global Const CS_NOCLOSE = &H200

CS_NOKEYCVT
Global Const CS_NOKEYCVT = &H100

CS_OWNDC
Global Const CS_OWNDC = &H20

CS_PARENTDC
Global Const CS_PARENTDC = &H80

CS_SAVEBITS
Global Const CS_SAVEBITS = &H800

CS_VREDRAW
Global Const CS_VREDRAW = &H1

CTLCOLOR_BTN
Global Const CTLCOLOR_BTN = 3

CTLCOLOR_DLG
Global Const CTLCOLOR_DLG = 4

CTLCOLOR_EDIT
Global Const CTLCOLOR_EDIT = 1

CTLCOLOR_LISTBOX
Global Const CTLCOLOR_LISTBOX = 2

CTLCOLOR_MAX
Global Const CTLCOLOR_MAX = 8

CTLCOLOR_MSGBOX
Global Const CTLCOLOR_MSGBOX = 0

CTLCOLOR_SCROLLBAR
Global Const CTLCOLOR_SCROLLBAR = 5

CTLCOLOR_STATIC
Global Const CTLCOLOR_STATIC = 6

CURVECAPS
Global Const CURVECAPS = 28

CW_USEDEFAULT
Global Const CW_USEDEFAULT = &H8000

BF_APPLICATION
Global Const DBF_APPLICATION = &H0008

DBF_DRIVER
Global Const DBF_DRIVER = &H0010

DBF_ERROR
Global Const DBF_ERROR = &H8000

DBF_FATAL
Global Const DBF_FATAL = &Hc000

DBF_GDI
Global Const DBF_GDI = &H0400

DBF_KERNEL
Global Const DBF_KERNEL = &H1000

DBF_KRN_LOADMODULE
Global Const DBF_KRN_LOADMODULE = &H0002

DBF_KRN_MEMMAN
Global Const DBF_KRN_MEMMAN = &H0001

DBF_KRN_SEGMENTLOAD
Global Const DBF_KRN_SEGMENTLOAD = &H0004

DBF_MMSYSTEM
Global Const DBF_MMSYSTEM = &H0040

DBF_PENWIN
Global Const DBF_PENWIN = &H0020

DBF_TRACE
Global Const DBF_TRACE = &H0000

DBF_USER
Global Const DBF_USER = &H0800

DBF_WARNING
Global Const DBF_WARNING = &H4000

DBO_BUFFERFILL
Global Const DBO_BUFFERFILL = &H0004

DBO_CHECKFREE
Global Const DBO_CHECKFREE = &H0020

DBO_CHECKHEAP
Global Const DBO_CHECKHEAP = &H0001

DBO_DISABLEGPTRAPPING
Global Const DBO_DISABLEGPTRAPPING = &H0010

DBO_INT3BREAK
Global Const DBO_INT3BREAK = &H0100

DBO_NOERRORBREAK
Global Const DBO_NOERRORBREAK = &H0800

DBO_NOFATALBREAK
Global Const DBO_NOFATALBREAK = &H0400

DBO_SILENT
Global Const DBO_SILENT = &H8000

DBO_TRACEBREAK
Global Const DBO_TRACEBREAK = &H2000

DBO_WARNINGBREAK
Global Const DBO_WARNINGBREAK = &H1000

DC_HASDEFID
Global Const DC_HASDEFID = &H534%

DCB_ACCUMULATE
Global Const DCB_ACCUMULATE = &H0002

DCB_DIRTY
Global Const DCB_DIRTY = DCB_ACCUMULATE

DCB_DISABLE
Global Const DCB_DISABLE = &H0008

DCB_ENABLE
Global Const DCB_ENABLE = &H0004

DCB_RESET
Global Const DCB_RESET = &H0001

DCB_SET
Global Const DCB_SET = (DCB_RESET Or DCB_ACCUMULATE)

DCX_CACHE
Global Const DCX_CACHE = &H00000002&

DCX_CLIPCHILDREN
Global Const DCX_CLIPCHILDREN = &H00000008&

DCX_CLIPSIBLINGS
Global Const DCX_CLIPSIBLINGS = &H00000010&

DCX_EXCLUDERGN
Global Const DCX_EXCLUDERGN = &H00000040&

DCX_INTERSECTRGN
Global Const DCX_INTERSECTRGN = &H00000080&

DCX_LOCKWINDOWUPDATE
Global Const DCX_LOCKWINDOWUPDATE = &H00000400&

DCX_PARENTCLIP
Global Const DCX_PARENTCLIP = &H00000020&

DCX_USESTYLE
Global Const DCX_USESTYLE = &H00010000&

DCX_WINDOW
Global Const DCX_WINDOW = &H00000001&

DEFAULT_PALETTE
Global Const DEFAULT_PALETTE = 15

DEFAULT_PITCH
Global Const DEFAULT_PITCH = 0

DEFAULT_QUALITY
Global Const DEFAULT_QUALITY = 0

DEVICE_DEFAULT_FONT
Global Const DEVICE_DEFAULT_FONT = 14

DEVICEDATA
Global Const DEVICEDATA = 19

DIB_PAL_COLORS
Global Const DIB_PAL_COLORS = 1

DIB_RGB_COLORS
Global Const DIB_RGB_COLORS = 0

DKGRAY_BRUSH
Global Const DKGRAY_BRUSH = 3

DLGC_BUTTON
Global Const DLGC_BUTTON = &H2000

DLGC_DEFPUSHBUTTON
Global Const DLGC_DEFPUSHBUTTON = &H10

DLGC_HASSETSEL
Global Const DLGC_HASSETSEL = &H8

DLGC_RADIOBUTTON
Global Const DLGC_RADIOBUTTON = &H40

DLGC_STATIC
Global Const DLGC_STATIC = &H100

DLGC_UNDEFPUSHBUTTON
Global Const DLGC_UNDEFPUSHBUTTON = &H20

DLGC_WANTALLKEYS
Global Const DLGC_WANTALLKEYS = &H4

DLGC_WANTARROWS
Global Const DLGC_WANTARROWS = &H1

DLGC_WANTCHARS
Global Const DLGC_WANTCHARS = &H80

DLGC_WANTMESSAGE
Global Const DLGC_WANTMESSAGE = &H4

DLGC_WANTTAB
Global Const DLGC_WANTTAB = &H2

DLGWINDOWEXTRA
Global Const DLGWINDOWEXTRA = 30

DM_GETDEFID
Global Const DM_GETDEFID = WM_USER+0

DM_SETDEFID
Global Const DM_SETDEFID = WM_USER+1

DRAFT_QUALITY
Global Const DRAFT_QUALITY = 1

DRAFTMODE
Global Const DRAFTMODE = 7

DRAWPATTERNRECT
Global Const DRAWPATTERNRECT = 25

DRIVE_FIXED
Global Const DRIVE_FIXED = 3

DRIVE_REMOTE
Global Const DRIVE_REMOTE = 4

DRIVE_REMOVABLE
Global Const DRIVE_REMOVABLE = 2

DRIVERVERSION
Global Const DRIVERVERSION = 0

DRV_CLOSE
Global Const DRV_CLOSE = &H0004

DRV_CONFIGURE
Global Const DRV_CONFIGURE = &H0007

DRV_DISABLE
Global Const DRV_DISABLE = &H0005

DRV_ENABLE
Global Const DRV_ENABLE = &H0002

DRV_EXITAPPLICATION
Global Const DRV_EXITAPPLICATION = &H000C

DRV_EXITSESSION
Global Const DRV_EXITSESSION = &H000B

DRV_FREE
Global Const DRV_FREE = &H0006

DRV_INSTALL
Global Const DRV_INSTALL = &H0009

DRV_LOAD
Global Const DRV_LOAD = &H0001

DRV_OPEN
Global Const DRV_OPEN = &H0003

DRV_POWER
Global Const DRV_POWER = &H000F

DRV_QUERYCONFIGURE
Global Const DRV_QUERYCONFIGURE = &H0008

DRV_REMOVE
Global Const DRV_REMOVE = &H000A

DRV_RESERVED
Global Const DRV_RESERVED = &H0800

DRV_USER
Global Const DRV_USER = &H4000

DRVCNF_CANCEL
Global Const DRVCNF_CANCEL = &H0000

DRVCNF_OK
Global Const DRVCNF_OK = &H0001

DRVCNF_RESTART
Global Const DRVCNF_RESTART = &H0002

DRVEA_ABNORMALEXIT
Global Const DRVEA_ABNORMALEXIT = &H0002

DRVEA_NORMALEXIT
Global Const DRVEA_NORMALEXIT = &H0001

DS_ABSALIGN
Global Const DS_ABSALIGN = &H1&

DS_LOCALEDIT
Global Const DS_LOCALEDIT = &H20&

DS_MODALFRAME
Global Const DS_MODALFRAME = &H80&

DS_NOIDLEMSG
Global Const DS_NOIDLEMSG = &H100&

DS_SETFONT
Global Const DS_SETFONT = &H40&

DS_SYSMODAL
Global Const DS_SYSMODAL = &H2&

DSTINVERT
Global Const DSTINVERT = &H550009

DT_BOTTOM
Global Const DT_BOTTOM = &H8

DT_CALCRECT
Global Const DT_CALCRECT = &H400

DT_CENTER
Global Const DT_CENTER = &H1

DT_CHARSTREAM
Global Const DT_CHARSTREAM = 4

DT_DISPFILE
Global Const DT_DISPFILE = 6

DT_EXPANDTABS
Global Const DT_EXPANDTABS = &H40

DT_EXTERNALLEADING
Global Const DT_EXTERNALLEADING = &H200

DT_INTERNAL
Global Const DT_INTERNAL = &H1000

DT_LEFT
Global Const DT_LEFT = &H0

DT_METAFILE
Global Const DT_METAFILE = 5

DT_NOCLIP
Global Const DT_NOCLIP = &H100

DT_NOPREFIX
Global Const DT_NOPREFIX = &H800

DT_PLOTTER
Global Const DT_PLOTTER = 0

DT_RASCAMERA
Global Const DT_RASCAMERA = 3

DT_RASDISPLAY
Global Const DT_RASDISPLAY = 1

DT_RASPRINTER
Global Const DT_RASPRINTER = 2

DT_RIGHT
Global Const DT_RIGHT = &H2

DT_SINGLELINE
Global Const DT_SINGLELINE = &H20

DT_TABSTOP
Global Const DT_TABSTOP = &H80

DT_TOP
Global Const DT_TOP = &H0

DT_VCENTER
Global Const DT_VCENTER = &H4

DT_WORDBREAK
Global Const DT_WORDBREAK = &H10

M_CANUNDO
Global Const EM_CANUNDO = WM_USER+22

EM_EMPTYUNDOBUFFER
Global Const EM_EMPTYUNDOBUFFER = WM_USER+29

EM_FMTLINES
Global Const EM_FMTLINES = WM_USER+24

EM_GETFIRSTVISIBLELINE
Global Const EM_GETFIRSTVISIBLELINE = (WM_USER+30)

EM_GETHANDLE
Global Const EM_GETHANDLE = WM_USER+13

EM_GETLINE
Global Const EM_GETLINE = WM_USER+20

EM_GETLINECOUNT
Global Const EM_GETLINECOUNT = WM_USER+10

EM_GETMODIFY
Global Const EM_GETMODIFY = WM_USER+8

EM_GETPASSWORDCHAR
Global Const EM_GETPASSWORDCHAR = (WM_USER+34)

EM_GETRECT
Global Const EM_GETRECT = WM_USER+2

EM_GETSEL
Global Const EM_GETSEL = WM_USER+0

EM_GETTHUMB
Global Const EM_GETTHUMB = WM_USER+14

EM_GETWORDBREAKPROC
Global Const EM_GETWORDBREAKPROC = (WM_USER+33)

EM_LIMITTEXT
Global Const EM_LIMITTEXT = WM_USER+21

EM_LINEFROMCHAR
Global Const EM_LINEFROMCHAR = WM_USER+25

EM_LINEINDEX
Global Const EM_LINEINDEX = WM_USER+11

EM_LINELENGTH
Global Const EM_LINELENGTH = WM_USER+17

EM_LINESCROLL
Global Const EM_LINESCROLL = WM_USER+6

EM_MSGMAX
Global Const EM_MSGMAX = WM_USER+30

EM_REPLACESEL
Global Const EM_REPLACESEL = WM_USER+18

EM_SCROLL
Global Const EM_SCROLL = WM_USER+5

EM_SETFONT
Global Const EM_SETFONT = WM_USER+19

EM_SETHANDLE
Global Const EM_SETHANDLE = WM_USER+12

EM_SETMODIFY
Global Const EM_SETMODIFY = WM_USER+9

EM_SETPASSWORDCHAR
Global Const EM_SETPASSWORDCHAR = WM_USER+28

EM_SETREADONLY
Global Const EM_SETREADONLY = (WM_USER+31)

EM_SETRECT
Global Const EM_SETRECT = WM_USER+3

EM_SETRECTNP
Global Const EM_SETRECTNP = WM_USER+4

EM_SETSEL
Global Const EM_SETSEL = WM_USER+1

EM_SETTABSTOPS
Global Const EM_SETTABSTOPS = WM_USER+27

EM_SETWORDBREAK
Global Const EM_SETWORDBREAK = WM_USER+26

EM_SETWORDBREAKPROC
Global Const EM_SETWORDBREAKPROC = (WM_USER+32)

EM_UNDO
Global Const EM_UNDO = WM_USER+23

EN_CHANGE
Global Const EN_CHANGE = &H300

EN_ERRSPACE
Global Const EN_ERRSPACE = &H500

EN_HSCROLL
Global Const EN_HSCROLL = &H601

EN_KILLFOCUS
Global Const EN_KILLFOCUS = &H200

EN_MAXTEXT
Global Const EN_MAXTEXT = &H501

EN_SETFOCUS
Global Const EN_SETFOCUS = &H100

EN_UPDATE
Global Const EN_UPDATE = &H400

EN_VSCROLL
Global Const EN_VSCROLL = &H602

ENABLEDUPLEX
Global Const ENABLEDUPLEX = 28

ENABLEPAIRKERNING
Global Const ENABLEPAIRKERNING = 769

ENABLERELATIVEWIDTHS
Global Const ENABLERELATIVEWIDTHS = 768

END_PATH
Global Const END_PATH = 4098

ENDDOCAPI
Global Const ENDDOCAPI = 11

ENUMPAPERBINS
Global Const ENUMPAPERBINS = 31

ENUMPAPERMETRICS
Global Const ENUMPAPERMETRICS = 34

EPSPRINTING
Global Const EPSPRINTING = 33

ERR_ALLOCRES
Global Const ERR_ALLOCRES = &H0007

ERR_BAD_ATOM
Global Const ERR_BAD_ATOM = &H6024

ERR_BAD_CID
Global Const ERR_BAD_CID = &H6045

ERR_BAD_COORDS
Global Const ERR_BAD_COORDS = &H7060

ERR_BAD_DFLAGS
Global Const ERR_BAD_DFLAGS = &H7005

ERR_BAD_DINDEX
Global Const ERR_BAD_DINDEX = &H7006

ERR_BAD_DVALUE
Global Const ERR_BAD_DVALUE = &H7004

ERR_BAD_FLAGS
Global Const ERR_BAD_FLAGS = &H6002

ERR_BAD_FUNC_PTR
Global Const ERR_BAD_FUNC_PTR = &H7008

ERR_BAD_GDI_OBJECT
Global Const ERR_BAD_GDI_OBJECT = &H6061

ERR_BAD_GLOBAL_HANDLE
Global Const ERR_BAD_GLOBAL_HANDLE = &H6022

ERR_BAD_HANDLE
Global Const ERR_BAD_HANDLE = &H600b

ERR_BAD_HBITMAP
Global Const ERR_BAD_HBITMAP = &H6066

ERR_BAD_HBRUSH
Global Const ERR_BAD_HBRUSH = &H6065

ERR_BAD_HCURSOR
Global Const ERR_BAD_HCURSOR = &H6042

ERR_BAD_HDC
Global Const ERR_BAD_HDC = &H6062

ERR_BAD_HDRVR
Global Const ERR_BAD_HDRVR = &H6046

ERR_BAD_HDWP
Global Const ERR_BAD_HDWP = &H6044

ERR_BAD_HFILE
Global Const ERR_BAD_HFILE = &H6025

ERR_BAD_HFONT
Global Const ERR_BAD_HFONT = &H6064

ERR_BAD_HICON
Global Const ERR_BAD_HICON = &H6043

ERR_BAD_HINSTANCE
Global Const ERR_BAD_HINSTANCE = &H6020

ERR_BAD_HMENU
Global Const ERR_BAD_HMENU = &H6041

ERR_BAD_HMETAFILE
Global Const ERR_BAD_HMETAFILE = &H6069

ERR_BAD_HMODULE
Global Const ERR_BAD_HMODULE = &H6021

ERR_BAD_HPALETTE
Global Const ERR_BAD_HPALETTE = &H6068

ERR_BAD_HPEN
Global Const ERR_BAD_HPEN = &H6063

ERR_BAD_HRGN
Global Const ERR_BAD_HRGN = &H6067

ERR_BAD_HWND
Global Const ERR_BAD_HWND = &H6040

ERR_BAD_INDEX
Global Const ERR_BAD_INDEX = &H6003

ERR_BAD_LOCAL_HANDLE
Global Const ERR_BAD_LOCAL_HANDLE = &H6023

ERR_BAD_PTR
Global Const ERR_BAD_PTR = &H7007

ERR_BAD_SELECTOR
Global Const ERR_BAD_SELECTOR = &H6009

ERR_BAD_STRING_PTR
Global Const ERR_BAD_STRING_PTR = &H700a

ERR_BAD_VALUE
Global Const ERR_BAD_VALUE = &H6001

ERR_BADINDEX
Global Const ERR_BADINDEX = &H0049

ERR_BYTE
Global Const ERR_BYTE = &H1000

ERR_CREATEDC
Global Const ERR_CREATEDC = &H0080

ERR_CREATEDLG
Global Const ERR_CREATEDLG = &H0040

ERR_CREATEDLG2
Global Const ERR_CREATEDLG2 = &H0041

ERR_CREATEMENU
Global Const ERR_CREATEMENU = &H004a

ERR_CREATEMETA
Global Const ERR_CREATEMETA = &H0081

ERR_CREATEWND
Global Const ERR_CREATEWND = &H0044

ERR_DCBUSY
Global Const ERR_DCBUSY = &H0043

ERR_DELOBJSELECTED
Global Const ERR_DELOBJSELECTED = &H0082

ERR_DWORD
Global Const ERR_DWORD = &H3000

ERR_GALLOC
Global Const ERR_GALLOC = &H0001

ERR_GLOCK
Global Const ERR_GLOCK = &H0003

ERR_GREALLOC
Global Const ERR_GREALLOC = &H0002

ERR_LALLOC
Global Const ERR_LALLOC = &H0004

ERR_LLOCK
Global Const ERR_LLOCK = &H0006

ERR_LOADMENU
Global Const ERR_LOADMENU = &H0047

ERR_LOADMODULE
Global Const ERR_LOADMODULE = &H0009

ERR_LOADSTR
Global Const ERR_LOADSTR = &H0046

ERR_LOCKRES
Global Const ERR_LOCKRES = &H0008

ERR_LREALLOC
Global Const ERR_LREALLOC = &H0005

ERR_NESTEDBEGINPAINT
Global Const ERR_NESTEDBEGINPAINT = &H0048

ERR_PARAM
Global Const ERR_PARAM = &H4000

ERR_REGISTERCLASS
Global Const ERR_REGISTERCLASS = &H0042

ERR_SELBITMAP
Global Const ERR_SELBITMAP = &H0083

ERR_SIZE_MASK
Global Const ERR_SIZE_MASK = &H3000

ERR_STRUCEXTRA
Global Const ERR_STRUCEXTRA = &H0045

ERR_WARNING
Global Const ERR_WARNING = &H8000

ERR_WORD
Global Const ERR_WORD = &H2000

ERRORAPI
Global Const ERRORAPI = 0

ES_AUTOHSCROLL
Global Const ES_AUTOHSCROLL = &H80&

ES_AUTOVSCROLL
Global Const ES_AUTOVSCROLL = &H40&

ES_CENTER
Global Const ES_CENTER = &H1&

ES_LEFT
Global Const ES_LEFT = &H0&

ES_LOWERCASE
Global Const ES_LOWERCASE = &H10&

ES_MULTILINE
Global Const ES_MULTILINE = &H4&

ES_NOHIDESEL
Global Const ES_NOHIDESEL = &H100&

ES_OEMCONVERT
Global Const ES_OEMCONVERT = &H400&

ES_PASSWORD
Global Const ES_PASSWORD = &H20&

ES_READONLY
Global Const ES_READONLY = &H00000800&

ES_RIGHT
Global Const ES_RIGHT = &H2&

ES_UPPERCASE
Global Const ES_UPPERCASE = &H8&

ES_WANTRETURN
Global Const ES_WANTRETURN = &H00001000&

ETO_CLIPPED
Global Const ETO_CLIPPED = 4

ETO_GRAYED
Global Const ETO_GRAYED = 1

ETO_OPAQUE
Global Const ETO_OPAQUE = 2

EV_BREAK
Global Const EV_BREAK = &H40

EV_CTS
Global Const EV_CTS = &H8

EV_DSR
Global Const EV_DSR = &H10

EV_ERR
Global Const EV_ERR = &H80

EV_PERR
Global Const EV_PERR = &H200

EV_RING
Global Const EV_RING = &H100

EV_RLSD
Global Const EV_RLSD = &H20

EV_RXCHAR
Global Const EV_RXCHAR = &H1

EV_RXFLAG
Global Const EV_RXFLAG = &H2

EV_TXEMPTY
Global Const EV_TXEMPTY = &H4

EVENPARITY
Global Const EVENPARITY = 2

EW_REBOOTSYSTEM
Global Const EW_REBOOTSYSTEM = &H43

EXT_DEVICE_CAPS
Global Const EXT_DEVICE_CAPS = 4099

EXTTEXTOUTAPI
Global Const EXTTEXTOUTAPI = 512

F_DECORATIVE
Global Const FF_DECORATIVE = 80

FF_DONTCARE
Global Const FF_DONTCARE = 0

FF_MODERN
Global Const FF_MODERN = 48

FF_ROMAN
Global Const FF_ROMAN = 16

FF_SCRIPT
Global Const FF_SCRIPT = 64

FF_SWISS
Global Const FF_SWISS = 32

FIXED_PITCH
Global Const FIXED_PITCH = 1

FLOODFILLBORDER
Global Const FLOODFILLBORDER = 0

FLOODFILLSURFACE
Global Const FLOODFILLSURFACE = 1

FLUSHOUTPUT
Global Const FLUSHOUTPUT = 6

FW_BLACK
Global Const FW_BLACK = FW_HEAVY

FW_BOLD
Global Const FW_BOLD = 700

FW_DEMIBOLD
Global Const FW_DEMIBOLD = FW_SEMIBOLD

FW_DONTCARE
Global Const FW_DONTCARE = 0

FW_EXTRABOLD
Global Const FW_EXTRABOLD = 800

FW_EXTRALIGHT
Global Const FW_EXTRALIGHT = 200

FW_HEAVY
Global Const FW_HEAVY = 900

FW_LIGHT
Global Const FW_LIGHT = 300

FW_MEDIUM
Global Const FW_MEDIUM = 500

FW_NORMAL
Global Const FW_NORMAL = 400

FW_REGULAR
Global Const FW_REGULAR = FW_NORMAL

FW_SEMIBOLD
Global Const FW_SEMIBOLD = 600

FW_THIN
Global Const FW_THIN = 100

FW_ULTRABOLD
Global Const FW_ULTRABOLD = FW_EXTRABOLD

FW_ULTRALIGHT
Global Const FW_ULTRALIGHT = FW_EXTRALIGHT

PI CONSTANT (G ~ Q)

A B C D E F , **G H I J K L M N O P Q** , R S T U V W X Y Z

B References

CL_MENUNAME
Global Const GCL_MENUNAME = (-8)

GCL_WNDPROC
Global Const GCL_WNDPROC = (-24)

GCW_ATOM
Global Const GCW_ATOM = (-32)

GCW_CBCLSEXTRA
Global Const GCW_CBCLSEXTRA = (-20)

GCW_CBWNDEXTRA
Global Const GCW_CBWNDEXTRA = (-18)

GCW_HBRBACKGROUND
Global Const GCW_HBRBACKGROUND = (-10)

GCW_HCURSOR
Global Const GCW_HCURSOR = (-12)

GCW_HICON
Global Const GCW_HICON = (-14)

GCW_HMODULE
Global Const GCW_HMODULE = (-16)

GCW_STYLE
Global Const GCW_STYLE = (-26)

GETBASEIRQ
Global Const GETBASEIRQ = 10

GETCOLORTABLE
Global Const GETCOLORTABLE = 5

GETEXTENDEDTEXTMETRICS
Global Const GETEXTENDEDTEXT-METRICS = 256

GETEXTENTTABLE
Global Const GETEXTENTTABLE = 257

GETMAXCOM
Global Const GETMAXCOM = 9

GETMAXLPT
Global Const GETMAXLPT = 8

GETPAIRKERNTABLE
Global Const GETPAIRKERNTABLE = 258

GETPENWIDTH
Global Const GETPENWIDTH = 16

GETPHYSPAGESIZE
Global Const GETPHYSPAGESIZE = 12

GETPRINTINGOFFSET
Global Const GETPRINTINGOFFSET = 13

GETSCALINGFACTOR
Global Const GETSCALINGFACTOR = 14

GETSETPAPERBINS
Global Const GETSETPAPERBINS = 29

GETSETPAPERMETRICS
Global Const GETSETPAPERMETRICS = 35

GETSETPRINTORIENT
Global Const GETSETPRINTORIENT = 30

GETTECHNOLOGY
Global Const GETTECHNOLOGY = 20

GETTRACKKERNTABLE
Global Const GETTRACKKERNTABLE = 259

GETVECTORBRUSHSIZE
Global Const GETVECTORBRUSHSIZE = 27

GETVECTORPENSIZE
Global Const GETVECTORPENSIZE = 26

GFSR_GDIRESOURCES
Global Const GFSR_GDIRESOURCES = &H0001

GFSR_SYSTEMRESOURCES
Global Const GFSR_SYSTEMRESOURCES = &H0000

GFSR_USERRESOURCES
Global Const GFSR_USERRESOURCES = &H0002

GGO_BITMAP
Global Const GGO_BITMAP = 1

GGO_METRICS
Global Const GGO_METRICS = 0

GGO_NATIVE
Global Const GGO_NATIVE = 2

GHND
Global Const GHND = (GMEM_MOVEABLE Or GMEM_ZEROINIT)

GMEM_DDESHARE
Global Const GMEM_DDESHARE = &H2000

GMEM_DISCARDABLE
Global Const GMEM_DISCARDABLE = &H100

GMEM_DISCARDED
Global Const GMEM_DISCARDED = &H4000

GMEM_FIXED
Global Const GMEM_FIXED = &H0

GMEM_LOCKCOUNT
Global Const GMEM_LOCKCOUNT = &HFF

GMEM_LOWER
Global Const GMEM_LOWER = GMEM_NOT_BANKED

GMEM_MODIFY
Global Const GMEM_MODIFY = &H80

GMEM_MOVEABLE
Global Const GMEM_MOVEABLE = &H2

GMEM_NOCOMPACT
Global Const GMEM_NOCOMPACT = &H10

GMEM_NODISCARD
Global Const GMEM_NODISCARD = &H20

GMEM_NOT_BANKED
Global Const GMEM_NOT_BANKED = &H1000

GMEM_NOTIFY
Global Const GMEM_NOTIFY = &H4000

GMEM_SHARE
Global Const GMEM_SHARE = &H2000

GMEM_ZEROINIT
Global Const GMEM_ZEROINIT = &H40

GND_FIRSTINSTANCEONLY
Global Const GND_FIRSTINSTANCEONLY = &H00000001

GND_FORWARD
Global Const GND_FORWARD = &H00000000

GND_REVERSE
Global Const GND_REVERSE = &H00000002

GPTR
Global Const GPTR = (GMEM_FIXED Or GMEM_ZEROINIT)

GRAY_BRUSH
Global Const GRAY_BRUSH = 2

GW_CHILD
Global Const GW_CHILD = 5

GW_HWNDFIRST
Global Const GW_HWNDFIRST = 0

GW_HWNDLAST
Global Const GW_HWNDLAST = 1

GW_HWNDNEXT
Global Const GW_HWNDNEXT = 2

GW_HWNDPREV
Global Const GW_HWNDPREV = 3

GW_OWNER
Global Const GW_OWNER = 4

GWL_EXSTYLE
Global Const GWL_EXSTYLE = (-20)

GWL_STYLE
Global Const GWL_STYLE = (-16)

GWL_WNDPROC
Global Const GWL_WNDPROC = (-4)

GWW_HINSTANCE
Global Const GWW_HINSTANCE = (-6)

GWW_HWNDPARENT
Global Const GWW_HWNDPARENT = (-8)

GWW_ID
Global Const GWW_ID = (-12)

C_ACTION
Global Const HC_ACTION = 0

HC_GETNEXT
Global Const HC_GETNEXT = 1

HC_LPFNNEXT
Global Const HC_LPFNNEXT = (-1)

HC_LPLPFNNEXT
Global Const HC_LPLPFNNEXT = (-2)

HC_NOREM
Global Const HC_NOREM = 3

HC_NOREMOVE
Global Const HC_NOREMOVE = 3

HC_SKIP
Global Const HC_SKIP = 2

HC_SYSMODALOFF
Global Const HC_SYSMODALOFF = 5

HC_SYSMODALON
Global Const HC_SYSMODALON = 4

HCBT_MINMAX
Global Const HCBT_MINMAX = 1

HCBT_MOVESIZE
Global Const HCBT_MOVESIZE = 0

HCBT_QS
Global Const HCBT_QS = 2

HELP_COMMAND
Global Const HELP_COMMAND = &H102

HELP_CONTENTS
Global Const HELP_CONTENTS = &H3

HELP_CONTEXT
Global Const HELP_CONTEXT = &H1

HELP_CONTEXTPOPUP
Global Const HELP_CONTEXTPOPUP = &H8

HELP_FORCEFILE
Global Const HELP_FORCEFILE = &H9

HELP_HELPONHELP
Global Const HELP_HELPONHELP = &H4

HELP_INDEX
Global Const HELP_INDEX = &H3

HELP_KEY
Global Const HELP_KEY = &H101

HELP_MULTIKEY
Global Const HELP_MULTIKEY = &H201

HELP_PARTIALKEY
Global Const HELP_PARTIALKEY = &H105

HELP_QUIT
Global Const HELP_QUIT = &H2

HELP_SETCONTENTS
Global Const HELP_SETCONTENTS = &H5

HELP_SETINDEX
Global Const HELP_SETINDEX = &H5

HELP_SETWINPOS
Global Const HELP_SETWINPOS = &H203

HIDE_WINDOW
Global Const HIDE_WINDOW = 0

HOLLOW_BRUSH
Global Const HOLLOW_BRUSH = NULL_BRUSH

HORZRES
Global Const HORZRES = 8

HORZSIZE
Global Const HORZSIZE = 4

HS_BDIAGONAL
Global Const HS_BDIAGONAL = 3

HS_CROSS
Global Const HS_CROSS = 4

HS_DIAGCROSS
Global Const HS_DIAGCROSS = 5

HS_FDIAGONAL
Global Const HS_FDIAGONAL = 2

HS_HORIZONTAL
Global Const HS_HORIZONTAL = 0

HS_VERTICAL
Global Const HS_VERTICAL = 1

HSHELL_ACTIVATESHELLWINDOW
Global Const HSHELL_ACTIVATESHELLWINDOW = 3

HSHELL_WINDOWCREATED
Global Const HSHELL_WINDOWCREATED = 1

HSHELL_WINDOWDESTROYED
Global Const HSHELL_WINDOWDESTROYED = 2

HTBOTTOM
Global Const HTBOTTOM = 15

HTBOTTOMLEFT
Global Const HTBOTTOMLEFT = 16

HTBOTTOMRIGHT
Global Const HTBOTTOMRIGHT = 17

HTCAPTION
Global Const HTCAPTION = 2

HTCLIENT
Global Const HTCLIENT = 1

HTERROR
Global Const HTERROR = (-2)

HTGROWBOX
Global Const HTGROWBOX = 4

HTHSCROLL
Global Const HTHSCROLL = 6

HTLEFT
Global Const HTLEFT = 10

HTMENU
Global Const HTMENU = 5

HTNOWHERE
Global Const HTNOWHERE = 0

HTREDUCE
Global Const HTREDUCE = 8

HTRIGHT
Global Const HTRIGHT = 11

HTSIZE
Global Const HTSIZE = HTGROWBOX

HTSIZEFIRST
Global Const HTSIZEFIRST = HTLEFT

HTSIZELAST
Global Const HTSIZELAST = HTBOTTOMRIGHT

HTSYSMENU
Global Const HTSYSMENU = 3

HTTOP
Global Const HTTOP = 12

HTTOPLEFT
Global Const HTTOPLEFT = 13

HTTOPRIGHT
Global Const HTTOPRIGHT = 14

HTTRANSPARENT
Global Const HTTRANSPARENT = (-1)

HTVSCROLL
Global Const HTVSCROLL = 7

HTZOOM
Global Const HTZOOM = 9

HWND_BROADCAST
Global Const HWND_BROADCAST = &Hffff

HWND_DESKTOP
Global Const HWND_DESKTOP = 0

HWND_TOP
Global Const HWND_TOP = 0

HWND_BOTTOM
Global Const HWND_BOTTOM = 1

HWND_TOPMOST
Global Const HWND_TOPMOST = -1

HWND_NOTOPMOST
Global Const HWND_NOTOPMOST = -2

DABORT
Global Const IDABORT = 3

IDC_ARROW
Global Const IDC_ARROW = 32512&

IDC_CROSS
Global Const IDC_CROSS = 32515&

IDC_IBEAM
Global Const IDC_IBEAM = 32513&

IDC_ICON
Global Const IDC_ICON = 32641&

IDC_SIZE
Global Const IDC_SIZE = 32640&

IDC_SIZENESW
Global Const IDC_SIZENESW = 32643&

IDC_SIZENS
Global Const IDC_SIZENS = 32645&

IDC_SIZENWSE
Global Const IDC_SIZENWSE = 32642&

IDC_SIZEWE
Global Const IDC_SIZEWE = 32644&

IDC_UPARROW
Global Const IDC_UPARROW = 32516&

IDC_WAIT
Global Const IDC_WAIT = 32514&

IDCANCEL
Global Const IDCANCEL = 2

IDI_APPLICATION
Global Const IDI_APPLICATION = 32512&

IDI_ASTERISK
Global Const IDI_ASTERISK = 32516&

IDI_EXCLAMATION
Global Const IDI_EXCLAMATION = 32515&

IDI_HAND
Global Const IDI_HAND = 32513&

IDI_QUESTION
Global Const IDI_QUESTION = 32514&

IDIGNORE
Global Const IDIGNORE = 5

IDNO
Global Const IDNO = 7

IDOK
Global Const IDOK = 1

IDRETRY
Global Const IDRETRY = 4

IDYES
Global Const IDYES = 6

IE_BADID
Global Const IE_BADID = (-1)

IE_BAUDRATE
Global Const IE_BAUDRATE = (-12)

IE_BYTESIZE
Global Const IE_BYTESIZE = (-11)

IE_DEFAULT
Global Const IE_DEFAULT = (-5)

IE_HARDWARE
Global Const IE_HARDWARE = (-10)

IE_MEMORY
Global Const IE_MEMORY = (-4)

IE_NOPEN
Global Const IE_NOPEN = (-3)

IE_OPEN
Global Const IE_OPEN = (-2)

IGNORE
Global Const IGNORE = 0

INFINITE
Global Const INFINITE = &HFFFF

B_ADDSTRING
Global Const LB_ADDSTRING = (WM_USER+1)

LB_CTLCODE
Global Const LB_CTLCODE = 0&

LB_DELETESTRING
Global Const LB_DELETESTRING = (WM_USER+3)

LB_DIR
Global Const LB_DIR = (WM_USER+14)

LB_ERR
Global Const LB_ERR = (-1)

LB_ERRSPACE
Global Const LB_ERRSPACE = (-2)

LB_FINDSTRING
Global Const LB_FINDSTRING = (WM_USER+16)

LB_FINDSTRINGEXACT
Global Const LB_FINDSTRINGEXACT = (WM_USER+35)

LB_GETCOUNT
Global Const LB_GETCOUNT = (WM_USER+12)

LB_GETCURSEL
Global Const LB_GETCURSEL = (WM_USER+9)

LB_GETHORIZONTALEXTENT
Global Const LB_GETHORIZONTALEX-TENT = (WM_USER+20)

LB_GETITEMDATA
Global Const LB_GETITEMDATA = (WM_USER+26)

LB_GETITEMHEIGHT
Global Const LB_GETITEMHEIGHT = (WM_USER+34)

LB_GETITEMRECT
Global Const LB_GETITEMRECT = (WM_USER+25)

LB_GETSEL
Global Const LB_GETSEL = (WM_USER+8)

LB_GETSELCOUNT
Global Const LB_GETSELCOUNT = (WM_USER+17)

LB_GETSELITEMS
Global Const LB_GETSELITEMS = (WM_USER+18)

LB_GETTEXT
Global Const LB_GETTEXT = (WM_USER+10)

LB_GETTEXTLEN
Global Const LB_GETTEXTLEN = (WM_USER+11)

LB_GETTOPINDEX
Global Const LB_GETTOPINDEX = (WM_USER+15)

LB_INSERTSTRING
Global Const LB_INSERTSTRING = (WM_USER+2)

LB_MSGMAX
Global Const LB_MSGMAX = (WM_USER+33)

LB_OKAY
Global Const LB_OKAY = 0

LB_RESETCONTENT
Global Const LB_RESETCONTENT = (WM_USER+5)

LB_SELECTSTRING
Global Const LB_SELECTSTRING = (WM_USER+13)

LB_SELITEMRANGE
Global Const LB_SELITEMRANGE = (WM_USER+28)

LB_SETCOLUMNWIDTH
Global Const LB_SETCOLUMNWIDTH = (WM_USER+22)

LB_SETCURSEL
Global Const LB_SETCURSEL = (WM_USER+7)

LB_SETHORIZONTALEXTENT
Global Const LB_SETHORIZONTALEXTENT = (WM_USER+21)

LB_SETITEMDATA
Global Const LB_SETITEMDATA = (WM_USER+27)

LB_SETITEMHEIGHT
Global Const LB_SETITEMHEIGHT = (WM_USER+33)

LB_SETSEL
Global Const LB_SETSEL = (WM_USER+6)

LB_SETTABSTOPS
Global Const LB_SETTABSTOPS = (WM_USER+19)

LB_SETTOPINDEX
Global Const LB_SETTOPINDEX = (WM_USER+24)

LBN_DBLCLK
Global Const LBN_DBLCLK = 2

LBN_ERRSPACE
Global Const LBN_ERRSPACE = (-2)

LBN_KILLFOCUS
Global Const LBN_KILLFOCUS = 5

LBN_SELCANCEL
Global Const LBN_SELCANCEL = 3

LBN_SELCHANGE
Global Const LBN_SELCHANGE = 1

LBN_SETFOCUS
Global Const LBN_SETFOCUS = 4

LBS_DISABLENOSCROLL
Global Const LBS_DISABLENOSCROLL = &H1000&

LBS_EXTENDEDSEL
Global Const LBS_EXTENDEDSEL = &H800&

LBS_HASSTRINGS
Global Const LBS_HASSTRINGS = &H40&

LBS_MULTICOLUMN
Global Const LBS_MULTICOLUMN = &H200&

LBS_MULTIPLESEL
Global Const LBS_MULTIPLESEL = &H8&

LBS_NOINTEGRALHEIGHT
Global Const LBS_NOINTEGRALHEIGHT = &H100&

LBS_NOREDRAW
Global Const LBS_NOREDRAW = &H4&

LBS_NOTIFY
Global Const LBS_NOTIFY = &H1&

LBS_OWNERDRAWFIXED
Global Const LBS_OWNERDRAWFIXED = &H10&

LBS_OWNERDRAWVARIABLE
Global Const LBS_OWNERDRAWVARIABLE = &H20&

LBS_SORT
Global Const LBS_SORT = &H2&

LBS_STANDARD
Global Const LBS_STANDARD = (LBS_NOTIFY Or LBS_SORT Or WS_VSCROLL Or **WS_BORDER**)

LBS_USETABSTOPS
Global Const LBS_USETABSTOPS = &H80&

LBS_WANTKEYBOARDINPUT
Global Const LBS_WANTKEYBOARDINPUT = &H400&

LC_INTERIORS
Global Const LC_INTERIORS = 128

LC_MARKER
Global Const LC_MARKER = 4

LC_NONE
Global Const LC_NONE = 0

LC_POLYLINE
Global Const LC_POLYLINE = 2

LC_POLYMARKER
Global Const LC_POLYMARKER = 8

LC_STYLED
Global Const LC_STYLED = 32

LC_WIDE
Global Const LC_WIDE = 16

LC_WIDESTYLED
Global Const LC_WIDESTYLED = 64

LF_FACESIZE
Global Const LF_FACESIZE = 32

LF_FULLFACESIZE
Global Const LF_FULLFACESIZE = 64

LHND
Global Const LHND = (LMEM_MOVEABLE+LMEM_ZEROINIT)

LINECAPS
Global Const LINECAPS = 30

LMEM_DISCARDABLE
Global Const LMEM_DISCARDABLE = &HF00

LMEM_DISCARDED
Global Const LMEM_DISCARDED = &H4000

LMEM_FIXED
Global Const LMEM_FIXED = &H0

LMEM_LOCKCOUNT
Global Const LMEM_LOCKCOUNT = &HFF

LMEM_MODIFY
Global Const LMEM_MODIFY = &H80

LMEM_MOVEABLE
Global Const LMEM_MOVEABLE = &H2

LMEM_NOCOMPACT
Global Const LMEM_NOCOMPACT = &H10

LMEM_NODISCARD
Global Const LMEM_NODISCARD = &H20

LMEM_ZEROINIT
Global Const LMEM_ZEROINIT = &H40

LNOTIFY_DISCARD
Global Const LNOTIFY_DISCARD = 2

LNOTIFY_MOVE
Global Const LNOTIFY_MOVE = 1

LNOTIFY_OUTOFMEM
Global Const LNOTIFY_OUTOFMEM = 0

LOGPIXELSX
Global Const LOGPIXELSX = 88

LOGPIXELSY
Global Const LOGPIXELSY = 90

LPTR
Global Const LPTR = (LMEM_FIXED+LMEM_ZEROINIT)

LPTx
Global Const LPTx = &H80

LTGRAY_BRUSH
Global Const LTGRAY_BRUSH = 1

LZERROR_BADINHANDLE
Global Const LZERROR_BADINHANDLE = (-1)

LZERROR_BADOUTHANDLE
Global Const LZERROR_BADOUTHANDLE = (-2)

LZERROR_BADVALUE
Global Const LZERROR_BADVALUE = (-7)

LZERROR_GLOBALLOC
Global Const LZERROR_GLOBALLOC = (-5)

LZERROR_GLOBLOCK
Global Const LZERROR_GLOBLOCK = (-6)

LZERROR_READ
Global Const LZERROR_READ = (-3)

LZERROR_UNKNOWNALG
Global Const LZERROR_UNKNOWNALG = (-8)

LZERROR_WRITE
Global Const LZERROR_WRITE = (-4)

MA_ACTIVATE
Global Const MA_ACTIVATE = 1

MA_ACTIVATEANDEAT
Global Const MA_ACTIVATEANDEAT = 2

MA_NOACTIVATE
Global Const MA_NOACTIVATE = 3

MA_NOACTIVATEANDEAT
Global Const MA_NOACTIVATEANDEAT = 4

MARKPARITY
Global Const MARKPARITY = 3

MB_ABORTRETRYIGNORE
Global Const MB_ABORTRETRYIGNORE = &H2

MB_APPLMODAL
Global Const MB_APPLMODAL = &H0

MB_DEFBUTTON1
Global Const MB_DEFBUTTON1 = &H0

MB_DEFBUTTON2
Global Const MB_DEFBUTTON2 = &H100

MB_DEFBUTTON3
Global Const MB_DEFBUTTON3 = &H200

MB_DEFMASK
Global Const MB_DEFMASK = &HF00

MB_ICONASTERISK
Global Const MB_ICONASTERISK = &H40

MB_ICONEXCLAMATION
Global Const MB_ICONEXCLAMATION = &H30

MB_ICONHAND
Global Const MB_ICONHAND = &H10

MB_ICONINFORMATION
Global Const MB_ICONINFORMATION = MB_ICONASTERISK

MB_ICONMASK
Global Const MB_ICONMASK = &HF0

MB_ICONQUESTION
Global Const MB_ICONQUESTION = &H20

MB_ICONSTOP
Global Const MB_ICONSTOP = MB_ICONHAND

MB_MISCMASK
Global Const MB_MISCMASK = &HC000

MB_MODEMASK
Global Const MB_MODEMASK = &H3000

MB_NOFOCUS
Global Const MB_NOFOCUS = &H8000

MB_OK
Global Const MB_OK = &H0

MB_OKCANCEL
Global Const MB_OKCANCEL = &H1

MB_RETRYCANCEL
Global Const MB_RETRYCANCEL = &H5

MB_SYSTEMMODAL
Global Const MB_SYSTEMMODAL = &H1000

MB_TASKMODAL
Global Const MB_TASKMODAL = &H2000

MB_TYPEMASK
Global Const MB_TYPEMASK = &HF

MB_YESNO
Global Const MB_YESNO = &H4

MB_YESNOCANCEL
Global Const MB_YESNOCANCEL = &H3

MDIS_ALLCHILDSTYLES
Global Const MDIS_ALLCHILDSTYLES = &H0001

MDITILE_HORIZONTAL
Global Const MDITILE_HORIZONTAL = &H0001

MDITILE_SKIPDISABLED
Global Const MDITILE_SKIPDISABLED = &H0002

MDITILE_VERTICAL
Global Const MDITILE_VERTICAL = &H0000

MERGECOPY
Global Const MERGECOPY = &HC000CA

MERGEPAINT
Global Const MERGEPAINT = &HBB0226

META_ANIMATEPALETTE
Global Const META_ANIMATEPALETTE = &H436

META_ARC
Global Const META_ARC = &H817

META_BITBLT
Global Const META_BITBLT = &H922

META_CHORD
Global Const META_CHORD = &H830

META_CREATEBITMAP
Global Const META_CREATEBITMAP = &H6FE

META_CREATEBITMAPINDIRECT
Global Const META_CREATEBITMAPINDIRECT = &H2FD

META_CREATEBRUSH
Global Const META_CREATEBRUSH = &HF8

META_CREATEBRUSHINDIRECT
Global Const META_CREATEBRUSHINDIRECT = &H2FC

META_CREATEFONTINDIRECT
Global Const META_CREATEFONTINDIRECT = &H2FB

META_CREATEPALETTE
Global Const META_CREATEPALETTE = &Hf7

META_CREATEPATTERNBRUSH
Global Const META_CREATEPATTERNBRUSH = &H1F9

META_CREATEPENINDIRECT
Global Const META_CREATEPENINDIRECT = &H2FA

META_CREATEREGION
Global Const META_CREATEREGION = &H6FF

META_DELETEOBJECT
Global Const META_DELETEOBJECT = &H1f0

META_DIBBITBLT
Global Const META_DIBBITBLT = &H940

META_DIBCREATEPATTERNBRUSH
Global Const META_DIBCREATEPATTERNBRUSH = &H142

META_DIBSTRETCHBLT
Global Const META_DIBSTRETCHBLT = &Hb41

META_DRAWTEXT
Global Const META_DRAWTEXT = &H62F

META_ELLIPSE
Global Const META_ELLIPSE = &H418

META_ESCAPE
Global Const META_ESCAPE = &H626

META_EXCLUDECLIPRECT
Global Const META_EXCLUDECLIPRECT = &H415

META_EXTTEXTOUT
Global Const META_EXTTEXTOUT = &Ha32

META_FILLREGION
Global Const META_FILLREGION = &H228

META_FLOODFILL
Global Const META_FLOODFILL = &H419

META_FRAMEREGION
Global Const META_FRAMEREGION = &H429

META_INTERSECTCLIPRECT
Global Const META_INTERSECTCLIPRECT = &H416

META_INVERTREGION
Global Const META_INVERTREGION = &H12A

META_LINETO
Global Const META_LINETO = &H213

META_MOVETO
Global Const META_MOVETO = &H214

META_OFFSETCLIPRGN
Global Const META_OFFSETCLIPRGN = &H220

META_OFFSETVIEWPORTORG
Global Const META_OFFSETVIEWPORTORG = &H211

META_OFFSETWINDOWORG
Global Const META_OFFSETWINDOWORG = &H20F

META_PAINTREGION
Global Const META_PAINTREGION = &H12B

META_PATBLT
Global Const META_PATBLT = &H61D

META_PIE
Global Const META_PIE = &H81A

META_POLYGON
Global Const META_POLYGON = &H324

META_POLYLINE
Global Const META_POLYLINE = &H325

META_POLYPOLYGON
Global Const META_POLYPOLYGON = &H538

META_REALIZEPALETTE
Global Const META_REALIZEPALETTE = &H35

META_RECTANGLE
Global Const META_RECTANGLE = &H41B

META_RESIZEPALETTE
Global Const META_RESIZEPALETTE = &H139

META_RESTOREDC
Global Const META_RESTOREDC = &H127

META_ROUNDRECT
Global Const META_ROUNDRECT = &H61C

META_SAVEDC
Global Const META_SAVEDC = &H1E

META_SCALEVIEWPORTEXT
Global Const META_SCALEVIEWPORTEXT = &H412

META_SCALEWINDOWEXT
Global Const META_SCALEWINDOWEXT = &H400

META_SELECTCLIPREGION
Global Const META_SELECTCLIPREGION = &H12C

META_SELECTOBJECT
Global Const META_SELECTOBJECT = &H12D

META_SELECTPALETTE
Global Const META_SELECTPALETTE = &H234

META_SETBKCOLOR
Global Const META_SETBKCOLOR = &H201

META_SETBKMODE
Global Const META_SETBKMODE = &H102

META_SETDIBTODEV
Global Const META_SETDIBTODEV = &Hd33

META_SETMAPMODE
Global Const META_SETMAPMODE = &H103

META_SETMAPPERFLAGS
Global Const META_SETMAPPERFLAGS = &H231

META_SETPALENTRIES
Global Const META_SETPALENTRIES = &H37

META_SETPIXEL
Global Const META_SETPIXEL = &H41F

META_SETPOLYFILLMODE
Global Const META_SETPOLYFILLMODE = &H106

META_SETRELABS
Global Const META_SETRELABS = &H105

META_SETROP2
Global Const META_SETROP2 = &H104

META_SETSTRETCHBLTMODE
Global Const META_SETSTRETCHBLTMODE = &H107

META_SETTEXTALIGN
Global Const META_SETTEXTALIGN = &H12E

META_SETTEXTCHAREXTRA
Global Const META_SETTEXTCHAREXTRA = &H108

META_SETTEXTCOLOR
Global Const META_SETTEXTCOLOR = &H209

META_SETTEXTJUSTIFICATION
Global Const META_SETTEXTJUSTIFICATION = &H20A

META_SETVIEWPORTEXT
Global Const META_SETVIEWPORTEXT = &H20E

META_SETVIEWPORTORG
Global Const META_SETVIEWPORTORG = &H20D

META_SETWINDOWEXT
Global Const META_SETWINDOWEXT = &H20C

META_SETWINDOWORG
Global Const META_SETWINDOWORG = &H20B

META_STRETCHBLT
Global Const META_STRETCHBLT = &HB23

META_STRETCHDIB
Global Const META_STRETCHDIB = &Hf43

META_TEXTOUT
Global Const META_TEXTOUT = &H521

MF_APPEND
Global Const MF_APPEND = &H100

MF_BITMAP
Global Const MF_BITMAP = &H4

MF_BYCOMMAND
Global Const MF_BYCOMMAND = &H0

MF_BYPOSITION
Global Const MF_BYPOSITION = &H400

MF_CHANGE
Global Const MF_CHANGE = &H80

MF_CHECKED
Global Const MF_CHECKED = &H8

MF_DELETE
Global Const MF_DELETE = &H200

MF_DISABLED
Global Const MF_DISABLED = &H2

MF_ENABLED
Global Const MF_ENABLED = &H0

MF_END
Global Const MF_END = &H80

MF_GRAYED
Global Const MF_GRAYED = &H1

MF_HELP
Global Const MF_HELP = &H4000

MF_HILITE
Global Const MF_HILITE = &H80

MF_INSERT
Global Const MF_INSERT = &H0

MF_MENUBARBREAK
Global Const MF_MENUBARBREAK = &H20

MF_MENUBREAK
Global Const MF_MENUBREAK = &H40

MF_MOUSESELECT
Global Const MF_MOUSESELECT = &H8000

MF_OWNERDRAW
Global Const MF_OWNERDRAW = &H100

MF_POPUP
Global Const MF_POPUP = &H10

MF_REMOVE
Global Const MF_REMOVE = &H1000

MF_SEPARATOR
Global Const MF_SEPARATOR = &H800

MF_STRING
Global Const MF_STRING = &H0

MF_SYSMENU
Global Const MF_SYSMENU = &H2000

MF_UNCHECKED
Global Const MF_UNCHECKED = &H0

MF_UNHILITE
Global Const MF_UNHILITE = &H0

MF_USECHECKBITMAPS
Global Const MF_USECHECKBITMAPS = &H200

MFCOMMENT
Global Const MFCOMMENT = 15

MK_CONTROL
Global Const MK_CONTROL = &H8

MK_LBUTTON
Global Const MK_LBUTTON = &H1

MK_MBUTTON
Global Const MK_MBUTTON = &H10

MK_RBUTTON
Global Const MK_RBUTTON = &H2

MK_SHIFT
Global Const MK_SHIFT = &H4

MM_ANISOTROPIC
Global Const MM_ANISOTROPIC = 8

MM_HIENGLISH
Global Const MM_HIENGLISH = 5

MM_HIMETRIC
Global Const MM_HIMETRIC = 3

MM_ISOTROPIC
Global Const MM_ISOTROPIC = 7

MM_LOENGLISH
Global Const MM_LOENGLISH = 4

MM_LOMETRIC
Global Const MM_LOMETRIC = 2

MM_TEXT
Global Const MM_TEXT = 1

MM_TWIPS
Global Const MM_TWIPS = 6

MSGF_DIALOGBOX
Global Const MSGF_DIALOGBOX = 0

MSGF_MENU
Global Const MSGF_MENU = 2

MSGF_MESSAGEBOX
Global Const MSGF_MESSAGEBOX = 1

MSGF_MOVE
Global Const MSGF_MOVE = 3

MSGF_NEXTWINDOW
Global Const MSGF_NEXTWINDOW = 6

MSGF_SCROLLBAR
Global Const MSGF_SCROLLBAR = 5

MSGF_SIZE
Global Const MSGF_SIZE = 4

Appendix H

API CONSTANTS N THROUGH Z

NEXTBAND
Global Const NEXTBAND = 3

NONZEROLHND
Global Const NONZEROLHND = (LMEM_MOVEABLE)

NONZEROLPTR
Global Const NONZEROLPTR = (LMEM_FIXED)

NOPARITY
Global Const NOPARITY = 0

NOTSRCCOPY
Global Const NOTSRCCOPY = &H330008

NOTSRCERASE
Global Const NOTSRCERASE = &H1100A6

NTM_BOLD
Global Const NTM_BOLD = &H00000020&

NTM_ITALIC
Global Const NTM_ITALIC = &H00000001&

NTM_REGULAR
Global Const NTM_REGULAR = &H00000040&

NULL_BRUSH
Global Const NULL_BRUSH = 5

NULL_PEN
Global Const NULL_PEN = 8

NULLREGION
Global Const NULLREGION = 1

NUMBRUSHES
Global Const NUMBRUSHES = 16

NUMCOLORS
Global Const NUMCOLORS = 24

NUMFONTS
Global Const NUMFONTS = 22

NUMMARKERS
Global Const NUMMARKERS = 20

NUMPENS
Global Const NUMPENS = 18

NUMRESERVED
Global Const NUMRESERVED = 106

BM_BTNCORNERS
Global Const OBM_BTNCORNERS = 32758

OBM_BTSIZE
Global Const OBM_BTSIZE = 32761

OBM_CHECK
Global Const OBM_CHECK = 32760

OBM_CHECKBOXES
Global Const OBM_CHECKBOXES = 32759

OBM_CLOSE
Global Const OBM_CLOSE = 32754

OBM_COMBO
Global Const OBM_COMBO = 32738

OBM_DNARROW
Global Const OBM_DNARROW = 32752

OBM_DNARROWD
Global Const OBM_DNARROWD = 32742

OBM_DNARROWI
Global Const OBM_DNARROWI = 32736

OBM_LFARROW
Global Const OBM_LFARROW = 32750

OBM_LFARROWD
Global Const OBM_LFARROWD = 32740

OBM_LFARROWI
Global Const OBM_LFARROWI = 32734

OBM_MNARROW
Global Const OBM_MNARROW = 32739

OBM_OLD_CLOSE
Global Const OBM_OLD_CLOSE = 32767

OBM_OLD_DNARROW
Global Const OBM_OLD_DNARROW = 32764

OBM_OLD_LFARROW
Global Const OBM_OLD_LFARROW = 32762

OBM_OLD_REDUCE
Global Const OBM_OLD_REDUCE = 32757

OBM_OLD_RESTORE
Global Const OBM_OLD_RESTORE = 32755

OBM_OLD_RGARROW
Global Const OBM_OLD_RGARROW = 32763

OBM_OLD_UPARROW
Global Const OBM_OLD_UPARROW = 32765

OBM_OLD_ZOOM
Global Const OBM_OLD_ZOOM = 32756

OBM_REDUCE
Global Const OBM_REDUCE = 32749

OBM_REDUCED
Global Const OBM_REDUCED = 32746

OBM_RESTORE
Global Const OBM_RESTORE = 32747

OBM_RESTORED
Global Const OBM_RESTORED = 32744

OBM_RGARROW
Global Const OBM_RGARROW = 32751

OBM_RGARROWD
Global Const OBM_RGARROWD = 32741

OBM_RGARROWI
Global Const OBM_RGARROWI = 32735

OBM_SIZE
Global Const OBM_SIZE = 32766

OBM_UPARROW
Global Const OBM_UPARROW = 32753

OBM_UPARROWD
Global Const OBM_UPARROWD = 32743

OBM_UPARROWI
Global Const OBM_UPARROWI = 32737

OBM_ZOOM
Global Const OBM_ZOOM = 32748

OBM_ZOOMD
Global Const OBM_ZOOMD = 32745

OCR_CROSS
Global Const OCR_CROSS = 32515

OCR_IBEAM
Global Const OCR_IBEAM = 32513

OCR_ICOCUR
Global Const OCR_ICOCUR = 32647

OCR_ICON
Global Const OCR_ICON = 32641

OCR_NORMAL
Global Const OCR_NORMAL = 32512

OCR_SIZE
Global Const OCR_SIZE = 32640

OCR_SIZEALL
Global Const OCR_SIZEALL = 32646

OCR_SIZENESW
Global Const OCR_SIZENESW = 32643

OCR_SIZENS
Global Const OCR_SIZENS = 32645

OCR_SIZENWSE
Global Const OCR_SIZENWSE = 32642

OCR_SIZEWE
Global Const OCR_SIZEWE = 32644

OCR_UP
Global Const OCR_UP = 32516

OCR_WAIT
Global Const OCR_WAIT = 32514

ODA_DRAWENTIRE
Global Const ODA_DRAWENTIRE = &H1

ODA_FOCUS
Global Const ODA_FOCUS = &H4

ODA_SELECT
Global Const ODA_SELECT = &H2

ODDPARITY
Global Const ODDPARITY = 1

ODS_CHECKED
Global Const ODS_CHECKED = &H8

ODS_DISABLED
Global Const ODS_DISABLED = &H4

ODS_FOCUS
Global Const ODS_FOCUS = &H10

ODS_GRAYED
Global Const ODS_GRAYED = &H2

ODS_SELECTED
Global Const ODS_SELECTED = &H1

ODT_BUTTON
Global Const ODT_BUTTON = 4

ODT_COMBOBOX
Global Const ODT_COMBOBOX = 3

ODT_LISTBOX
Global Const ODT_LISTBOX = 2

ODT_MENU
Global Const ODT_MENU = 1

OEM_CHARSET
Global Const OEM_CHARSET = 255

OEM_FIXED_FONT
Global Const OEM_FIXED_FONT = 10

OF_CANCEL
Global Const OF_CANCEL = &H800

OF_CREATE
Global Const OF_CREATE = &H1000

OF_DELETE
Global Const OF_DELETE = &H200

OF_EXIST
Global Const OF_EXIST = &H4000

OF_PARSE
Global Const OF_PARSE = &H100

OF_PROMPT
Global Const OF_PROMPT = &H2000

OF_READ
Global Const OF_READ = &H0

OF_READWRITE
Global Const OF_READWRITE = &H2

OF_REOPEN
Global Const OF_REOPEN = &H8000

OF_SHARE_COMPAT
Global Const OF_SHARE_COMPAT = &H0

OF_SHARE_DENY_NONE
Global Const OF_SHARE_DENY_NONE = &H40

OF_SHARE_DENY_READ
Global Const OF_SHARE_DENY_READ = &H30

OF_SHARE_DENY_WRITE
Global Const OF_SHARE_DENY_WRITE = &H20

OF_SHARE_EXCLUSIVE
Global Const OF_SHARE_EXCLUSIVE = &H10

OF_VERIFY
Global Const OF_VERIFY = &H400

OF_WRITE
Global Const OF_WRITE = &H1

OIC_BANG
Global Const OIC_BANG = 32515

OIC_HAND
Global Const OIC_HAND = 32513

OIC_NOTE
Global Const OIC_NOTE = 32516

OIC_QUES
Global Const OIC_QUES = 32514

OIC_SAMPLE
Global Const OIC_SAMPLE = 32512

ONE5STOPBITS
Global Const ONE5STOPBITS = 1

ONESTOPBIT
Global Const ONESTOPBIT = 0

OPAQUE
Global Const OPAQUE = 2

ORD_LANGDRIVER
Global Const ORD_LANGDRIVER = 1

OUT_CHARACTER_PRECIS
Global Const OUT_CHARACTER_PRECIS = 2

OUT_DEFAULT_PRECIS
Global Const OUT_DEFAULT_PRECIS = 0

OUT_DEVICE_PRECIS
Global Const OUT_DEVICE_PRECIS = 5

OUT_RASTER_PRECIS
Global Const OUT_RASTER_PRECIS = 6

OUT_STRING_PRECIS
Global Const OUT_STRING_PRECIS = 1

OUT_STROKE_PRECIS
Global Const OUT_STROKE_PRECIS = 3

OUT_TT_ONLY_PRECIS
Global Const OUT_TT_ONLY_PRECIS = 7

OUT_TT_PRECIS
Global Const OUT_TT_PRECIS = 4

PASSTHROUGH
Global Const PASSTHROUGH = 19

PATCOPY
Global Const PATCOPY = &HF00021

PATINVERT
Global Const PATINVERT = &H5A0049

PATPAINT
Global Const PATPAINT = &HFB0A09

PC_EXPLICIT
Global Const PC_EXPLICIT = &H2

PC_INTERIORS
Global Const PC_INTERIORS = 128

PC_NOCOLLAPSE
Global Const PC_NOCOLLAPSE = &H4

PC_NONE
Global Const PC_NONE = 0

PC_POLYGON
Global Const PC_POLYGON = 1

PC_RECTANGLE
Global Const PC_RECTANGLE = 2

PC_RESERVED
Global Const PC_RESERVED = &H1

PC_SCANLINE
Global Const PC_SCANLINE = 8

PC_STYLED
Global Const PC_STYLED = 32

PC_TRAPEZOID
Global Const PC_TRAPEZOID = 4

PC_WIDE
Global Const PC_WIDE = 16

PC_WIDESTYLED
Global Const PC_WIDESTYLED = 64

PC_WINDPOLYGON
Global Const PC_WINDPOLYGON = 4

PDEVICESIZE
Global Const PDEVICESIZE = 26

PLANES
Global Const PLANES = 14

PM_NOREMOVE
Global Const PM_NOREMOVE = &H0

PM_NOYIELD
Global Const PM_NOYIELD = &H2

PM_REMOVE
Global Const PM_REMOVE = &H1

POLYGONALCAPS
Global Const POLYGONALCAPS = 32

POSTSCRIPT_DATA
Global Const POSTSCRIPT_DATA = 37

POSTSCRIPT_IGNORE
Global Const POSTSCRIPT_IGNORE = 38

PR_JOBSTATUS
Global Const PR_JOBSTATUS = &H0

PROOF_QUALITY
Global Const PROOF_QUALITY = 2

PS_DASH
Global Const PS_DASH = 1

PS_DASHDOT
Global Const PS_DASHDOT = 3

PS_DASHDOTDOT
Global Const PS_DASHDOTDOT = 4

PS_DOT
Global Const PS_DOT = 2

PS_INSIDEFRAME
Global Const PS_INSIDEFRAME = 6

PS_NULL
Global Const PS_NULL = 5

PS_SOLID
Global Const PS_SOLID = 0

PWR_CRITICALRESUME
Global Const PWR_CRITICALRESUME = 3

PWR_FAIL
Global Const PWR_FAIL = -1

PWR_OK
Global Const PWR_OK = 1

PWR_SUSPENDREQUEST
Global Const PWR_SUSPENDREQUEST = 1

PWR_SUSPENDRESUME
Global Const PWR_SUSPENDRESUME = 2

S_ALLINPUT
Global Const QS_ALLINPUT = &H007f

QS_KEY
Global Const QS_KEY = &H0001

QS_MOUSE
Global Const QS_MOUSE = (QS_MOUSEMOVE Or QS_MOUSEBUTTON)

QS_MOUSEBUTTON
Global Const QS_MOUSEBUTTON = &H0004

QS_MOUSEMOVE
Global Const QS_MOUSEMOVE = &H0002

QS_PAINT
Global Const QS_PAINT = &H0020

QS_POSTMESSAGE
Global Const QS_POSTMESSAGE = &H0008

QS_SENDMESSAGE
Global Const QS_SENDMESSAGE = &H0040

QS_TIMER
Global Const QS_TIMER = &H0010

QUERYESCSUPPORT
Global Const QUERYESCSUPPORT = 8

PI CONSTANT (R ~ Z)

A B C D E F , G H I J K L M N O P Q , R S T U V W X Y Z

B References

2_BLACK
Global Const R2_BLACK = 1

R2_COPYPEN
Global Const R2_COPYPEN = 13

R2_MASKNOTPEN
Global Const R2_MASKNOTPEN = 3

R2_MASKPEN
Global Const R2_MASKPEN = 9

R2_MASKPENNOT
Global Const R2_MASKPENNOT = 5

R2_MERGENOTPEN
Global Const R2_MERGENOTPEN = 12

R2_MERGEPEN
Global Const R2_MERGEPEN = 15

R2_MERGEPENNOT
Global Const R2_MERGEPENNOT = 14

R2_NOP
Global Const R2_NOP = 11

R2_NOT
Global Const R2_NOT = 6

R2_NOTCOPYPEN
Global Const R2_NOTCOPYPEN = 4

R2_NOTMASKPEN
Global Const R2_NOTMASKPEN = 8

R2_NOTMERGEPEN
Global Const R2_NOTMERGEPEN = 2

R2_NOTXORPEN
Global Const R2_NOTXORPEN = 10

R2_WHITE
Global Const R2_WHITE = 16

R2_XORPEN
Global Const R2_XORPEN = 7

RASTERCAPS
Global Const RASTERCAPS = 38

RC_BANDING
Global Const RC_BANDING = 2

RC_BIGFONT
Global Const RC_BIGFONT = &H400

RC_BITBLT
Global Const RC_BITBLT = 1

RC_BITMAP64
Global Const RC_BITMAP64 = 8

RC_DI_BITMAP
Global Const RC_DI_BITMAP = &H80

RC_DIBTODEV
Global Const RC_DIBTODEV = &H200

RC_FLOODFILL
Global Const RC_FLOODFILL = &H1000

RC_GDI20_OUTPUT
Global Const RC_GDI20_OUTPUT = &H10

RC_PALETTE
Global Const RC_PALETTE = &H100

RC_SCALING
Global Const RC_SCALING = 4

RC_STRETCHBLT
Global Const RC_STRETCHBLT = &H800

RC_STRETCHDIB
Global Const RC_STRETCHDIB = &H2000

RDW_ALLCHILDREN
Global Const RDW_ALLCHILDREN = &H0080

RDW_ERASE
Global Const RDW_ERASE = &H0004

RDW_ERASENOW
Global Const RDW_ERASENOW = &H0200

RDW_FRAME
Global Const RDW_FRAME = &H0400

RDW_INTERNALPAINT
Global Const RDW_INTERNALPAINT = &H0002

RDW_INVALIDATE
Global Const RDW_INVALIDATE = &H0001

RDW_NOCHILDREN
Global Const RDW_NOCHILDREN = &H0040

RDW_NOERASE
Global Const RDW_NOERASE = &H0020

RDW_NOFRAME
Global Const RDW_NOFRAME = &H0800

RDW_NOINTERNALPAINT
Global Const RDW_NOINTERNALPAINT = &H0010

RDW_UPDATENOW
Global Const RDW_UPDATENOW = &H0100

RDW_VALIDATE
Global Const RDW_VALIDATE = &H0008

READ_WRITE
Global Const READ_WRITE = 2

READAPI
Global Const READAPI = 0

RELATIVE
Global Const RELATIVE = 2

RESETDEV
Global Const RESETDEV = 7

RESTORE_CTM
Global Const RESTORE_CTM = 4100

RGN_AND
Global Const RGN_AND = 1

RGN_COPY
Global Const RGN_COPY = 5

RGN_DIFF
Global Const RGN_DIFF = 4

RGN_OR
Global Const RGN_OR = 2

RGN_XOR
Global Const RGN_XOR = 3

RT_ACCELERATOR
Global Const RT_ACCELERATOR = 9&

RT_BITMAP
Global Const RT_BITMAP = 2&

RT_CURSOR
Global Const RT_CURSOR = 1&

RT_DIALOG
Global Const RT_DIALOG = 5&

RT_FONT
Global Const RT_FONT = 8&

RT_FONTDIR
Global Const RT_FONTDIR = 7&

RT_ICON
Global Const RT_ICON = 3&

RT_MENU
Global Const RT_MENU = 4&

RT_RCDATA
Global Const RT_RCDATA = 10&

RT_STRING
Global Const RT_STRING = 6&

_ALLTHRESHOLD
Global Const S_ALLTHRESHOLD = 2

S_LEGATO
Global Const S_LEGATO = 1

S_NORMAL
Global Const S_NORMAL = 0

S_PERIOD1024
Global Const S_PERIOD1024 = 1

S_PERIOD2048
Global Const S_PERIOD2048 = 2

S_PERIOD512
Global Const S_PERIOD512 = 0

S_PERIODVOICE
Global Const S_PERIODVOICE = 3

S_QUEUEEMPTY
Global Const S_QUEUEEMPTY = 0

S_SERBDNT
Global Const S_SERBDNT = (-5)

S_SERDCC
Global Const S_SERDCC = (-7)

S_SERDDR
Global Const S_SERDDR = (-14)

S_SERDFQ
Global Const S_SERDFQ = (-13)

S_SERDLN
Global Const S_SERDLN = (-6)

S_SERDMD
Global Const S_SERDMD = (-10)

S_SERDPT
Global Const S_SERDPT = (-12)

S_SERDSH
Global Const S_SERDSH = (-11)

S_SERDSR
Global Const S_SERDSR = (-15)

S_SERDST
Global Const S_SERDST = (-16)

S_SERDTP
Global Const S_SERDTP = (-8)

S_SERDVL
Global Const S_SERDVL = (-9)

S_SERDVNA
Global Const S_SERDVNA = (-1)

S_SERMACT
Global Const S_SERMACT = (-3)

S_SEROFM
Global Const S_SEROFM = (-2)

S_SERQFUL
Global Const S_SERQFUL = (-4)

S_STACCATO
Global Const S_STACCATO = 2

S_THRESHOLD
Global Const S_THRESHOLD = 1

S_WHITE1024
Global Const S_WHITE1024 = 5

S_WHITE2048
Global Const S_WHITE2048 = 6

S_WHITE512
Global Const S_WHITE512 = 4

S_WHITEVOICE
Global Const S_WHITEVOICE = 7

SAVE_CTM
Global Const SAVE_CTM = 4101

SB_BOTH
Global Const SB_BOTH = 3

SB_BOTTOM
Global Const SB_BOTTOM = 7

SB_CTL
Global Const SB_CTL = 2

SB_ENDSCROLL
Global Const SB_ENDSCROLL = 8

SB_HORZ
Global Const SB_HORZ = 0

SB_LINEDOWN
Global Const SB_LINEDOWN = 1

SB_LINEUP
Global Const SB_LINEUP = 0

SB_PAGEDOWN
Global Const SB_PAGEDOWN = 3

SB_PAGEUP
Global Const SB_PAGEUP = 2

SB_THUMBPOSITION
Global Const SB_THUMBPOSITION = 4

SB_THUMBTRACK
Global Const SB_THUMBTRACK = 5

SB_TOP
Global Const SB_TOP = 6

SB_VERT
Global Const SB_VERT = 1

SBS_BOTTOMALIGN
Global Const SBS_BOTTOMALIGN = &H4&

SBS_HORZ
Global Const SBS_HORZ = &H0&

SBS_LEFTALIGN
Global Const SBS_LEFTALIGN = &H2&

SBS_RIGHTALIGN
Global Const SBS_RIGHTALIGN = &H4&

SBS_SIZEBOX
Global Const SBS_SIZEBOX = &H8&

SBS_SIZEBOXBOTTOMRIGHTALIGN
Global Const SBS_SIZEBOXBOTTOMRIGHTALIGN = &H4&

SBS_SIZEBOXTOPLEFTALIGN
Global Const SBS_SIZEBOXTOPLEFTALIGN = &H2&

SBS_TOPALIGN
Global Const SBS_TOPALIGN = &H2&

SBS_VERT
Global Const SBS_VERT = &H1&

SC_ARRANGE
Global Const SC_ARRANGE = &HF110

SC_CLOSE
Global Const SC_CLOSE = &HF060

SC_HSCROLL
Global Const SC_HSCROLL = &HF080

SC_ICON
Global Const SC_ICON = SC_MINIMIZE

SC_KEYMENU
Global Const SC_KEYMENU = &HF100

SC_MAXIMIZE
Global Const SC_MAXIMIZE = &HF030

SC_MINIMIZE
Global Const SC_MINIMIZE = &HF020

SC_MOUSEMENU
Global Const SC_MOUSEMENU = &HF090

SC_MOVE
Global Const SC_MOVE = &HF010

SC_NEXTWINDOW
Global Const SC_NEXTWINDOW = &HF040

SC_PREVWINDOW
Global Const SC_PREVWINDOW = &HF050

SC_RESTORE
Global Const SC_RESTORE = &HF120

SC_SIZE
Global Const SC_SIZE = &HF000

SC_TASKLIST
Global Const SC_TASKLIST = &HF130

SC_VSCROLL
Global Const SC_VSCROLL = &HF070

SC_ZOOM
Global Const SC_ZOOM = SC_MAXIMIZE

SDS_DIALOG
Global Const SDS_DIALOG = &H0008

SDS_MENU
Global Const SDS_MENU = &H0001

SDS_NOTASKQUEUE
Global Const SDS_NOTASKQUEUE = &H0004

SDS_SYSMODAL
Global Const SDS_SYSMODAL = &H0002

SDS_TASKLOCKED
Global Const SDS_TASKLOCKED = &H0010

SELECTPAPERSOURCE
Global Const SELECTPAPERSOURCE = 18

SET_ARC_DIRECTION
Global Const SET_ARC_DIRECTION = 4102

SET_BACKGROUND_COLOR
Global Const SET_BACKGROUND_COLOR = 4103

SET_BOUNDS
Global Const SET_BOUNDS = 4109

SET_CLIP_BOX
Global Const SET_CLIP_BOX = 4108

SET_MIRROR_MODE
Global Const SET_MIRROR_MODE = 4110

SET_POLY_MODE
Global Const SET_POLY_MODE = 4104

SET_SCREEN_ANGLE
Global Const SET_SCREEN_ANGLE = 4105

SET_SPREAD
Global Const SET_SPREAD = 4106

SETABORTPROC
Global Const SETABORTPROC = 9

SETALLJUSTVALUES
Global Const SETALLJUSTVALUES = 771

SETCHARSET
Global Const SETCHARSET = 772

SETCOLORTABLE
Global Const SETCOLORTABLE = 4

SETCOPYCOUNT
Global Const SETCOPYCOUNT = 17

SETDIBSCALING
Global Const SETDIBSCALING = 32

SETDTR
Global Const SETDTR = 5

SETENDCAP
Global Const SETENDCAP = 21

SETKERNTRACK
Global Const SETKERNTRACK = 770

SETLINEJOIN
Global Const SETLINEJOIN = 22

SETMITERLIMIT
Global Const SETMITERLIMIT = 23

SETRTS
Global Const SETRTS = 3

SETXOFF
Global Const SETXOFF = 1

SETXON
Global Const SETXON = 2

SHIFTJIS_CHARSET
Global Const SHIFTJIS_CHARSET = 128

SHOW_FULLSCREEN
Global Const SHOW_FULLSCREEN = 3

SHOW_ICONWINDOW
Global Const SHOW_ICONWINDOW = 2

SHOW_OPENNOACTIVATE
Global Const SHOW_OPENNOACTIVATE = 4

SHOW_OPENWINDOW
Global Const SHOW_OPENWINDOW = 1

SIMPLEREGION
Global Const SIMPLEREGION = 2

SIZEFULLSCREEN
Global Const SIZEFULLSCREEN = 2

SIZEICONIC
Global Const SIZEICONIC = 1

SIZENORMAL
Global Const SIZENORMAL = 0

SIZEPALETTE
Global Const SIZEPALETTE = 104

SIZEZOOMHIDE
Global Const SIZEZOOMHIDE = 4

SIZEZOOMSHOW
Global Const SIZEZOOMSHOW = 3

SM_CMETRICS
Global Const SM_CMETRICS = 36

SM_CXBORDER
Global Const SM_CXBORDER = 5

SM_CXCURSOR
Global Const SM_CXCURSOR = 13

SM_CXDLGFRAME
Global Const SM_CXDLGFRAME = 7

SM_CXDOUBLECLK
Global Const SM_CXDOUBLECLK = 36

SM_CXFRAME
Global Const SM_CXFRAME = 32

SM_CXFULLSCREEN
Global Const SM_CXFULLSCREEN = 16

SM_CXHSCROLL
Global Const SM_CXHSCROLL = 21

SM_CXHTHUMB
Global Const SM_CXHTHUMB = 10

SM_CXICON
Global Const SM_CXICON = 11

SM_CXICONSPACING
Global Const SM_CXICONSPACING = 38

SM_CXMIN
Global Const SM_CXMIN = 28

SM_CXMINTRACK
Global Const SM_CXMINTRACK = 34

SM_CXSCREEN
Global Const SM_CXSCREEN = 0

SM_CXSIZE
Global Const SM_CXSIZE = 30

SM_CXVSCROLL
Global Const SM_CXVSCROLL = 2

SM_CYBORDER
Global Const SM_CYBORDER = 6

SM_CYCAPTION
Global Const SM_CYCAPTION = 4

SM_CYCURSOR
Global Const SM_CYCURSOR = 14

SM_CYDLGFRAME
Global Const SM_CYDLGFRAME = 8

SM_CYDOUBLECLK
Global Const SM_CYDOUBLECLK = 37

SM_CYFRAME
Global Const SM_CYFRAME = 33

SM_CYFULLSCREEN
Global Const SM_CYFULLSCREEN = 17

SM_CYHSCROLL
Global Const SM_CYHSCROLL = 3

SM_CYICON
Global Const SM_CYICON = 12

SM_CYICONSPACING
Global Const SM_CYICONSPACING = 39

SM_CYKANJIWINDOW
Global Const SM_CYKANJIWINDOW = 18

SM_CYMENU
Global Const SM_CYMENU = 15

SM_CYMIN
Global Const SM_CYMIN = 29

SM_CYMINTRACK
Global Const SM_CYMINTRACK = 35

SM_CYSCREEN
Global Const SM_CYSCREEN = 1

SM_CYSIZE
Global Const SM_CYSIZE = 31

SM_CYVSCROLL
Global Const SM_CYVSCROLL = 20

SM_CYVTHUMB
Global Const SM_CYVTHUMB = 9

SM_DBCSENABLED
Global Const SM_DBCSENABLED = 42

SM_DEBUG
Global Const SM_DEBUG = 22

SM_MENUDROPALIGNMENT
Global Const SM_MENUDROPALIGNMENT = 40

SM_MOUSEPRESENT
Global Const SM_MOUSEPRESENT = 19

SM_PENWINDOWS
Global Const SM_PENWINDOWS = 41

SM_RESERVED1
Global Const SM_RESERVED1 = 24

SM_RESERVED2
Global Const SM_RESERVED2 = 25

SM_RESERVED3
Global Const SM_RESERVED3 = 26

SM_RESERVED4
Global Const SM_RESERVED4 = 27

SM_SWAPBUTTON
Global Const SM_SWAPBUTTON = 23

SP_APPABORT
Global Const SP_APPABORT = (-2)

SP_ERROR
Global Const SP_ERROR = (-1)

SP_NOTREPORTED
Global Const SP_NOTREPORTED = &H4000

SP_OUTOFDISK
Global Const SP_OUTOFDISK = (-4)

SP_OUTOFMEMORY
Global Const SP_OUTOFMEMORY = (-5)

SP_USERABORT
Global Const SP_USERABORT = (-3)

SPACEPARITY
Global Const SPACEPARITY = 4

SPI_GETBEEP
Global Const SPI_GETBEEP = 1

SPI_GETBORDER
Global Const SPI_GETBORDER = 5

SPI_GETFASTTASKSWITCH
Global Const SPI_GETFASTTASKSWITCH = 35

SPI_GETGRIDGRANULARITY
Global Const SPI_GETGRIDGRANULARITY = 18

SPI_GETICONTITLELOGFONT
Global Const SPI_GETICONTITLELOGFONT = 31

SPI_GETICONTITLEWRAP
Global Const SPI_GETICONTITLEWRAP = 25

SPI_GETKEYBOARDDELAY
Global Const SPI_GETKEYBOARDDELAY = 22

SPI_GETKEYBOARDSPEED
Global Const SPI_GETKEYBOARDSPEED = 10

SPI_GETMENUDROPALIGNMENT
Global Const SPI_GETMENUDROPALIGNMENT = 27

SPI_GETMOUSE
Global Const SPI_GETMOUSE = 3

SPI_GETSCREENSAVEACTIVE
Global Const SPI_GETSCREENSAVEACTIVE = 16

SPI_GETSCREENSAVETIMEOUT
Global Const SPI_GETSCREENSAVETIMEOUT = 14

SPI_ICONHORIZONTALSPACING
Global Const SPI_ICONHORIZONTALSPACING = 13

SPI_ICONVERTICALSPACING
Global Const SPI_ICONVERTICALSPACING = 24

SPI_LANGDRIVER
Global Const SPI_LANGDRIVER = 12

SPI_SETBEEP
Global Const SPI_SETBEEP = 2

SPI_SETBORDER
Global Const SPI_SETBORDER = 6

SPI_SETDESKPATTERN
Global Const SPI_SETDESKPATTERN = 21

SPI_SETDESKWALLPAPER
Global Const SPI_SETDESKWALLPAPER = 20

SPI_SETDOUBLECLICKTIME
Global Const SPI_SETDOUBLECLICKTIME = 32

SPI_SETDOUBLECLKHEIGHT
Global Const SPI_SETDOUBLECLKHEIGHT = 30

SPI_SETDOUBLECLKWIDTH
Global Const SPI_SETDOUBLECLKWIDTH = 29

SPI_SETFASTTASKSWITCH
Global Const SPI_SETFASTTASKSWITCH = 36

SPI_SETGRIDGRANULARITY
Global Const SPI_SETGRIDGRANULARITY = 19

SPI_SETICONTITLELOGFONT
Global Const SPI_SETICONTITLELOGFONT = 34

SPI_SETICONTITLEWRAP
Global Const SPI_SETICONTITLEWRAP = 26

SPI_SETKEYBOARDDELAY
Global Const SPI_SETKEYBOARDDELAY = 23

SPI_SETKEYBOARDSPEED
Global Const SPI_SETKEYBOARDSPEED = 11

SPI_SETMENUDROPALIGNMENT
Global Const SPI_SETMENUDROPALIGNMENT = 28

SPI_SETMOUSE
Global Const SPI_SETMOUSE = 4

SPI_SETMOUSEBUTTONSWAP
Global Const SPI_SETMOUSEBUTTONSWAP = 33

SPI_SETSCREENSAVEACTIVE
Global Const SPI_SETSCREENSAVEACTIVE = 17

SPI_SETSCREENSAVETIMEOUT
Global Const SPI_SETSCREENSAVETIMEOUT = 15

SPIF_SENDWININICHANGE
Global Const SPIF_SENDWININICHANGE = &H0002

SPIF_UPDATEINIFILE
Global Const SPIF_UPDATEINIFILE = &H0001

SRCAND
Global Const SRCAND = &H8800C6

SRCCOPY
Global Const SRCCOPY = &HCC0020

SRCERASE
Global Const SRCERASE = &H440328

SRCINVERT
Global Const SRCINVERT = &H660046

SRCPAINT
Global Const SRCPAINT = &HEE0086

SS_BLACKFRAME
Global Const SS_BLACKFRAME = &H7&

SS_BLACKRECT
Global Const SS_BLACKRECT = &H4&

SS_CENTER
Global Const SS_CENTER = &H1&

SS_GRAYFRAME
Global Const SS_GRAYFRAME = &H8&

SS_GRAYRECT
Global Const SS_GRAYRECT = &H5&

SS_ICON
Global Const SS_ICON = &H3&

SS_LEFT
Global Const SS_LEFT = &H0&

SS_LEFTNOWORDWRAP
Global Const SS_LEFTNOWORDWRAP = &HC&

SS_NOPREFIX
Global Const SS_NOPREFIX = &H80&

SS_RIGHT
Global Const SS_RIGHT = &H2&

SS_SIMPLE
Global Const SS_SIMPLE = &HB&

SS_USERITEM
Global Const SS_USERITEM = &HA&

SS_WHITEFRAME
Global Const SS_WHITEFRAME = &H9&

SS_WHITERECT
Global Const SS_WHITERECT = &H6&

ST_BEGINSWP
Global Const ST_BEGINSWP = 0

ST_ENDSWP
Global Const ST_ENDSWP = 1

STARTDOCAPI
Global Const STARTDOCAPI = 10

STM_GETICON
Global Const STM_GETICON = (WM_USER+1)

STM_SETICON
Global Const STM_SETICON = (WM_USER+0)

STRETCHBLTAPI
Global Const STRETCHBLTAPI = 2048

SW_ERASE
Global Const SW_ERASE = &H0004

SW_HIDE
Global Const SW_HIDE = 0

SW_INVALIDATE
Global Const SW_INVALIDATE = &H0002

SW_MAXIMIZE
Global Const SW_MAXIMIZE = 3

SW_MINIMIZE
Global Const SW_MINIMIZE = 6

SW_NORMAL
Global Const SW_NORMAL = 1

SW_OTHERUNZOOM
Global Const SW_OTHERUNZOOM = 4

SW_OTHERZOOM
Global Const SW_OTHERZOOM = 2

SW_PARENTCLOSING
Global Const SW_PARENTCLOSING = 1

SW_PARENTOPENING
Global Const SW_PARENTOPENING = 3

SW_RESTORE
Global Const SW_RESTORE = 9

SW_SCROLLCHILDREN
Global Const SW_SCROLLCHILDREN = &H0001

SW_SHOW
Global Const SW_SHOW = 5

SW_SHOWMAXIMIZED
Global Const SW_SHOWMAXIMIZED = 3

SW_SHOWMINIMIZED
Global Const SW_SHOWMINIMIZED = 2

SW_SHOWMINNOACTIVE
Global Const SW_SHOWMINNOACTIVE = 7

SW_SHOWNA
Global Const SW_SHOWNA = 8

SW_SHOWNOACTIVATE
Global Const SW_SHOWNOACTIVATE = 4

SW_SHOWNORMAL
Global Const SW_SHOWNORMAL = 1

SWP_DRAWFRAME
Global Const SWP_DRAWFRAME = &H20

SWP_HIDEWINDOW
Global Const SWP_HIDEWINDOW = &H80

SWP_NOACTIVATE
Global Const SWP_NOACTIVATE = &H10

SWP_NOCOPYBITS
Global Const SWP_NOCOPYBITS = &H100

SWP_NOMOVE
Global Const SWP_NOMOVE = &H2

SWP_NOREDRAW
Global Const SWP_NOREDRAW = &H8

SWP_NOREPOSITION
Global Const SWP_NOREPOSITION = &H200

SWP_NOSIZE
Global Const SWP_NOSIZE = &H1

SWP_NOZORDER
Global Const SWP_NOZORDER = &H4

SWP_SHOWWINDOW
Global Const SWP_SHOWWINDOW = &H40

SYMBOL_CHARSET
Global Const SYMBOL_CHARSET = 2

SYSPAL_NOSTATIC
Global Const SYSPAL_NOSTATIC = 2

SYSPAL_STATIC
Global Const SYSPAL_STATIC = 1

SYSTEM_FIXED_FONT
Global Const SYSTEM_FIXED_FONT = 16

SYSTEM_FONT
Global Const SYSTEM_FONT = 13

TA_BASELINE
Global Const TA_BASELINE = 24

TA_BOTTOM
Global Const TA_BOTTOM = 8

TA_CENTER
Global Const TA_CENTER = 6

TA_LEFT
Global Const TA_LEFT = 0

TA_NOUPDATECP
Global Const TA_NOUPDATECP = 0

TA_RIGHT
Global Const TA_RIGHT = 2

TA_TOP
Global Const TA_TOP = 0

TA_UPDATECP
Global Const TA_UPDATECP = 1

TC_CP_STROKE
Global Const TC_CP_STROKE = &H4

TC_CR_90
Global Const TC_CR_90 = &H8

TC_CR_ANY
Global Const TC_CR_ANY = &H10

TC_EA_DOUBLE
Global Const TC_EA_DOUBLE = &H200

TC_IA_ABLE
Global Const TC_IA_ABLE = &H400

TC_OP_CHARACTER
Global Const TC_OP_CHARACTER = &H1

TC_OP_STROKE
Global Const TC_OP_STROKE = &H2

TC_RA_ABLE
Global Const TC_RA_ABLE = &H2000

TC_RESERVED
Global Const TC_RESERVED = &H8000

TC_SA_CONTIN
Global Const TC_SA_CONTIN = &H100

TC_SA_DOUBLE
Global Const TC_SA_DOUBLE = &H40

TC_SA_INTEGER
Global Const TC_SA_INTEGER = &H80

TC_SF_X_YINDEP
Global Const TC_SF_X_YINDEP = &H20

TC_SO_ABLE
Global Const TC_SO_ABLE = &H1000

TC_UA_ABLE
Global Const TC_UA_ABLE = &H800

TC_VA_ABLE
Global Const TC_VA_ABLE = &H4000

TECHNOLOGY
Global Const TECHNOLOGY = 2

TEXTCAPS
Global Const TEXTCAPS = 34

TF_FORCEDRIVE
Global Const TF_FORCEDRIVE = &H80

TMPF_TRUETYPE
Global Const TMPF_TRUETYPE = &H04

TPM_CENTERALIGN
Global Const TPM_CENTERALIGN = &H0004

TPM_LEFTALIGN
Global Const TPM_LEFTALIGN = &H0000

TPM_RIGHTALIGN
Global Const TPM_RIGHTALIGN = &H0008

TPM_RIGHTBUTTON
Global Const TPM_RIGHTBUTTON = &H0002

TRANSFORM_CTM
Global Const TRANSFORM_CTM = 4107

TRANSPARENT
Global Const TRANSPARENT = 1

TRUETYPE_FONTTYPE
Global Const TRUETYPE_FONTTYPE = &H0004

TT_AVAILABLE
Global Const TT_AVAILABLE = &H0001

TT_ENABLED
Global Const TT_ENABLED = &H0002

TT_POLYGON_TYPE
Global Const TT_POLYGON_TYPE = 24

TT_PRIM_LINE
Global Const TT_PRIM_LINE = 1

TT_PRIM_QSPLINE
Global Const TT_PRIM_QSPLINE = 2

TWOSTOPBITS
Global Const TWOSTOPBITS = 2

VARIABLE_PITCH
Global Const VARIABLE_PITCH = 2

VERTRES
Global Const VERTRES = 10

VERTSIZE
Global Const VERTSIZE = 6

VK_ADD
Global Const VK_ADD = &H6B

VK_BACK
Global Const VK_BACK = &H8

VK_CANCEL
Global Const VK_CANCEL = &H3

VK_CAPITAL
Global Const VK_CAPITAL = &H14

VK_CLEAR
Global Const VK_CLEAR = &HC

VK_CONTROL
Global Const VK_CONTROL = &H11

VK_COPY
Global Const VK_COPY = &H2C not used by keyboards.

VK_DECIMAL
Global Const VK_DECIMAL = &H6E

VK_DELETE
Global Const VK_DELETE = &H2E

VK_DIVIDE
Global Const VK_DIVIDE = &H6F

VK_DOWN
Global Const VK_DOWN = &H28

VK_END
Global Const VK_END = &H23

VK_ESCAPE
Global Const VK_ESCAPE = &H1B

VK_EXECUTE
Global Const VK_EXECUTE = &H2B

VK_F1
Global Const VK_F1 = &H70

VK_F10
Global Const VK_F10 = &H79

VK_F11
Global Const VK_F11 = &H7A

VK_F12
Global Const VK_F12 = &H7B

VK_F13
Global Const VK_F13 = &H7C

VK_F14
Global Const VK_F14 = &H7D

VK_F15
Global Const VK_F15 = &H7E

VK_F16
Global Const VK_F16 = &H7F

VK_F2
Global Const VK_F2 = &H71

VK_F3
Global Const VK_F3 = &H72

VK_F4
Global Const VK_F4 = &H73

VK_F5
Global Const VK_F5 = &H74

VK_F6
Global Const VK_F6 = &H75

VK_F7
Global Const VK_F7 = &H76

VK_F8
Global Const VK_F8 = &H77

VK_F9
Global Const VK_F9 = &H78

VK_HELP
Global Const VK_HELP = &H2F

VK_HOME
Global Const VK_HOME = &H24

VK_INSERT
Global Const VK_INSERT = &H2D

VK_LBUTTON
Global Const VK_LBUTTON = &H1

VK_LEFT
Global Const VK_LEFT = &H25

VK_MBUTTON
Global Const VK_MBUTTON = &H4

VK_MENU
Global Const VK_MENU = &H12

VK_MULTIPLY
Global Const VK_MULTIPLY = &H6A

VK_NEXT
Global Const VK_NEXT = &H22

VK_NUMLOCK
Global Const VK_NUMLOCK = &H90

VK_NUMPAD0
Global Const VK_NUMPAD0 = &H60

VK_NUMPAD1
Global Const VK_NUMPAD1 = &H61

VK_NUMPAD2
Global Const VK_NUMPAD2 = &H62

VK_NUMPAD3
Global Const VK_NUMPAD3 = &H63

VK_NUMPAD4
Global Const VK_NUMPAD4 = &H64

VK_NUMPAD5
Global Const VK_NUMPAD5 = &H65

VK_NUMPAD6
Global Const VK_NUMPAD6 = &H66

VK_NUMPAD7
Global Const VK_NUMPAD7 = &H67

VK_NUMPAD8
Global Const VK_NUMPAD8 = &H68

VK_NUMPAD9
Global Const VK_NUMPAD9 = &H69

VK_PAUSE
Global Const VK_PAUSE = &H13

VK_PRINT
Global Const VK_PRINT = &H2A

VK_PRIOR
Global Const VK_PRIOR = &H21

VK_RBUTTON
Global Const VK_RBUTTON = &H2

VK_RETURN
Global Const VK_RETURN = &HD

VK_RIGHT
Global Const VK_RIGHT = &H27

VK_SELECT
Global Const VK_SELECT = &H29

VK_SEPARATOR
Global Const VK_SEPARATOR = &H6C

VK_SHIFT
Global Const VK_SHIFT = &H10

VK_SNAPSHOT
Global Const VK_SNAPSHOT = &H2C

VK_SPACE
Global Const VK_SPACE = &H20

VK_SUBTRACT
Global Const VK_SUBTRACT = &H6D

VK_TAB
Global Const VK_TAB = &H9

VK_UP
Global Const VK_UP = &H26

B_ISDELIMITER
Global Const WB_ISDELIMITER = 2

WB_LEFT
Global Const WB_LEFT = 0

WB_RIGHT
Global Const WB_RIGHT = 1

WC_DEFWINDOWPROC
Global Const WC_DEFWINDOWPROC = 3

WC_DRAWCAPTION
Global Const WC_DRAWCAPTION = 7

WC_INIT
Global Const WC_INIT = 1

WC_MINMAX
Global Const WC_MINMAX = 4

WC_MOVE
Global Const WC_MOVE = 5

WC_SIZE
Global Const WC_SIZE = 6

WC_SWP
Global Const WC_SWP = 2

WDI_ALLOCBREAK
Global Const WDI_ALLOCBREAK = &H0004

WDI_FILTER
Global Const WDI_FILTER = &H0002

WDI_OPTIONS
Global Const WDI_OPTIONS = &H0001

WEP_FREE_DLL
Global Const WEP_FREE_DLL = 0

WEP_SYSTEM_EXIT
Global Const WEP_SYSTEM_EXIT = 1

WF_80x87
Global Const WF_80x87 = &H400

WF_CPU086
Global Const WF_CPU086 = &H40

WF_CPU186
Global Const WF_CPU186 = &H80

WF_CPU286
Global Const WF_CPU286 = &H2

WF_CPU386
Global Const WF_CPU386 = &H4

WF_CPU486
Global Const WF_CPU486 = &H8

WF_ENHANCED
Global Const WF_ENHANCED = &H20

WF_LARGEFRAME
Global Const WF_LARGEFRAME = &H100

WF_PMODE
Global Const WF_PMODE = &H1

WF_SMALLFRAME
Global Const WF_SMALLFRAME = &H200

WF_STANDARD
Global Const WF_STANDARD = &H10

WF_WIN286
Global Const WF_WIN286 = &H10

WF_WIN386
Global Const WF_WIN386 = &H20

WH_CALLWNDPROC
Global Const WH_CALLWNDPROC = 4

WH_CBT
Global Const WH_CBT = 5

WH_DEBUG
Global Const WH_DEBUG = 9

WH_GETMESSAGE
Global Const WH_GETMESSAGE = 3

WH_HARDWARE
Global Const WH_HARDWARE = 8

WH_JOURNALPLAYBACK
Global Const WH_JOURNALPLAYBACK = 1

WH_JOURNALRECORD
Global Const WH_JOURNALRECORD = 0

WH_KEYBOARD
Global Const WH_KEYBOARD = 2

WH_MSGFILTER
Global Const WH_MSGFILTER = (-1)

WH_SHELL
Global Const WH_SHELL = 10

WH_SYSMSGFILTER
Global Const WH_SYSMSGFILTER = 6

WH_WINDOWMGR
Global Const WH_WINDOWMGR = 7

WHITE_BRUSH
Global Const WHITE_BRUSH = 0

WHITE_PEN
Global Const WHITE_PEN = 6

WHITENESS
Global Const WHITENESS = &HFF0062

WHITEONBLACK
Global Const WHITEONBLACK = 2

WINDING
Global Const WINDING = 2

WM_ACTIVATE
Global Const WM_ACTIVATE = &H6

WM_ACTIVATEAPP
Global Const WM_ACTIVATEAPP = &H1C

WM_ASKCBFORMATNAME
Global Const WM_ASKCBFORMATNAME = &H30C

WM_CANCELMODE
Global Const WM_CANCELMODE = &H1F

WM_CHANGECBCHAIN
Global Const WM_CHANGECBCHAIN = &H30D

WM_CHAR
Global Const WM_CHAR = &H102

WM_CHARTOITEM
Global Const WM_CHARTOITEM = &H2F

WM_CHILDACTIVATE
Global Const WM_CHILDACTIVATE = &H22

WM_CLEAR
Global Const WM_CLEAR = &H303

WM_CLOSE
Global Const WM_CLOSE = &H10

WM_COMMAND
Global Const WM_COMMAND = &H111

WM_COMMNOTIFY
Global Const WM_COMMNOTIFY = &H0044

WM_COMPACTING
Global Const WM_COMPACTING = &H41

WM_COMPAREITEM
Global Const WM_COMPAREITEM = &H39

WM_COPY
Global Const WM_COPY = &H301

WM_CREATE
Global Const WM_CREATE = &H1

WM_CTLCOLOR
Global Const WM_CTLCOLOR = &H19

WM_CUT
Global Const WM_CUT = &H300

WM_DEADCHAR
Global Const WM_DEADCHAR = &H103

WM_DELETEITEM
Global Const WM_DELETEITEM = &H2D

WM_DESTROY
Global Const WM_DESTROY = &H2

WM_DESTROYCLIPBOARD
Global Const WM_DESTROYCLIPBOARD = &H307

WM_DEVMODECHANGE
Global Const WM_DEVMODECHANGE = &H1B

WM_DRAWCLIPBOARD
Global Const WM_DRAWCLIPBOARD = &H308

WM_DRAWITEM
Global Const WM_DRAWITEM = &H2B

WM_ENABLE
Global Const WM_ENABLE = &HA

WM_ENDSESSION
Global Const WM_ENDSESSION = &H16

WM_ENTERIDLE
Global Const WM_ENTERIDLE = &H121

WM_ERASEBKGND
Global Const WM_ERASEBKGND = &H14

WM_FONTCHANGE
Global Const WM_FONTCHANGE = &H1D

WM_GETDLGCODE
Global Const WM_GETDLGCODE = &H87

WM_GETFONT
Global Const WM_GETFONT = &H31

WM_GETMINMAXINFO
Global Const WM_GETMINMAXINFO = &H24

WM_GETTEXT
Global Const WM_GETTEXT = &HD

WM_GETTEXTLENGTH
Global Const WM_GETTEXTLENGTH = &HE

WM_HSCROLL
Global Const WM_HSCROLL = &H114

WM_HSCROLLCLIPBOARD
Global Const WM_HSCROLLCLIPBOARD = &H30E

WM_ICONERASEBKGND
Global Const WM_ICONERASEBKGND = &H27

WM_INITDIALOG
Global Const WM_INITDIALOG = &H110

WM_INITMENU
Global Const WM_INITMENU = &H116

WM_INITMENUPOPUP
Global Const WM_INITMENUPOPUP = &H117

WM_KEYDOWN
Global Const WM_KEYDOWN = &H100

WM_KEYFIRST
Global Const WM_KEYFIRST = &H100

WM_KEYLAST
Global Const WM_KEYLAST = &H108

WM_KEYUP
Global Const WM_KEYUP = &H101

WM_KILLFOCUS
Global Const WM_KILLFOCUS = &H8

WM_LBUTTONDBLCLK
Global Const WM_LBUTTONDBLCLK = &H203

WM_LBUTTONDOWN
Global Const WM_LBUTTONDOWN = &H201

WM_LBUTTONUP
Global Const WM_LBUTTONUP = &H202

WM_MBUTTONDBLCLK
Global Const WM_MBUTTONDBLCLK = &H209

WM_MBUTTONDOWN
Global Const WM_MBUTTONDOWN = &H207

WM_MBUTTONUP
Global Const WM_MBUTTONUP = &H208

WM_MDIACTIVATE
Global Const WM_MDIACTIVATE = &H222

WM_MDICASCADE
Global Const WM_MDICASCADE = &H227

WM_MDICREATE
Global Const WM_MDICREATE = &H220

WM_MDIDESTROY
Global Const WM_MDIDESTROY = &H221

WM_MDIGETACTIVE
Global Const WM_MDIGETACTIVE = &H229

WM_MDIICONARRANGE
Global Const WM_MDIICONARRANGE = &H228

WM_MDIMAXIMIZE
Global Const WM_MDIMAXIMIZE = &H225

WM_MDINEXT
Global Const WM_MDINEXT = &H224

WM_MDIRESTORE
Global Const WM_MDIRESTORE = &H223

WM_MDISETMENU
Global Const WM_MDISETMENU = &H230

WM_MDITILE
Global Const WM_MDITILE = &H226

WM_MEASUREITEM
Global Const WM_MEASUREITEM = &H2C

WM_MENUCHAR
Global Const WM_MENUCHAR = &H120

WM_MENUSELECT
Global Const WM_MENUSELECT = &H11F

WM_MOUSEACTIVATE
Global Const WM_MOUSEACTIVATE = &H21

WM_MOUSEFIRST
Global Const WM_MOUSEFIRST = &H200

WM_MOUSELAST
Global Const WM_MOUSELAST = &H209

WM_MOUSEMOVE
Global Const WM_MOUSEMOVE = &H200

WM_MOVE
Global Const WM_MOVE = &H3

WM_NCACTIVATE
Global Const WM_NCACTIVATE = &H86

WM_NCCALCSIZE
Global Const WM_NCCALCSIZE = &H83

WM_NCCREATE
Global Const WM_NCCREATE = &H81

WM_NCDESTROY
Global Const WM_NCDESTROY = &H82

WM_NCHITTEST
Global Const WM_NCHITTEST = &H84

WM_NCLBUTTONDBLCLK
Global Const WM_NCLBUTTONDBLCLK = &HA3

WM_NCLBUTTONDOWN
Global Const WM_NCLBUTTONDOWN = &HA1

WM_NCLBUTTONUP
Global Const WM_NCLBUTTONUP = &HA2

WM_NCMBUTTONDBLCLK
Global Const WM_NCMBUTTONDBLCLK = &HA9

WM_NCMBUTTONDOWN
Global Const WM_NCMBUTTONDOWN = &HA7

WM_NCMBUTTONUP
Global Const WM_NCMBUTTONUP = &HA8

WM_NCMOUSEMOVE
Global Const WM_NCMOUSEMOVE = &HA0

WM_NCPAINT
Global Const WM_NCPAINT = &H85

WM_NCRBUTTONDBLCLK
Global Const WM_NCRBUTTONDBLCLK = &HA6

WM_NCRBUTTONDOWN
Global Const WM_NCRBUTTONDOWN = &HA4

WM_NCRBUTTONUP
Global Const WM_NCRBUTTONUP = &HA5

WM_NEXTDLGCTL
Global Const WM_NEXTDLGCTL = &H28

WM_NULL
Global Const WM_NULL = &H0

WM_PAINT
Global Const WM_PAINT = &HF

WM_PAINTCLIPBOARD
Global Const WM_PAINTCLIPBOARD = &H309

WM_PAINTICON
Global Const WM_PAINTICON = &H26

WM_PALETTECHANGED
Global Const WM_PALETTECHANGED = &H311

WM_PALETTEISCHANGING
Global Const WM_PALETTEISCHANGING = &H310

WM_PARENTNOTIFY
Global Const WM_PARENTNOTIFY = &H210

WM_PASTE
Global Const WM_PASTE = &H302

WM_POWER
Global Const WM_POWER = &H0048

WM_QUERYDRAGICON
Global Const WM_QUERYDRAGICON = &H37

WM_QUERYENDSESSION
Global Const WM_QUERYENDSESSION = &H11

WM_QUERYNEWPALETTE
Global Const WM_QUERYNEWPALETTE = &H30F

WM_QUERYOPEN
Global Const WM_QUERYOPEN = &H13

WM_QUEUESYNC
Global Const WM_QUEUESYNC = &H23

WM_QUIT
Global Const WM_QUIT = &H12

WM_RBUTTONDBLCLK
Global Const WM_RBUTTONDBLCLK = &H206

WM_RBUTTONDOWN
Global Const WM_RBUTTONDOWN = &H204

WM_RBUTTONUP
Global Const WM_RBUTTONUP = &H205

WM_RENDERALLFORMATS
Global Const WM_RENDERALLFORMATS = &H306

WM_RENDERFORMAT
Global Const WM_RENDERFORMAT = &H305

WM_SETCURSOR
Global Const WM_SETCURSOR = &H20

WM_SETFOCUS
Global Const WM_SETFOCUS = &H7

WM_SETFONT
Global Const WM_SETFONT = &H30

WM_SETREDRAW
Global Const WM_SETREDRAW = &HB

WM_SETTEXT
Global Const WM_SETTEXT = &HC

WM_SHOWWINDOW
Global Const WM_SHOWWINDOW = &H18

WM_SIZE
Global Const WM_SIZE = &H5

WM_SIZECLIPBOARD
Global Const WM_SIZECLIPBOARD = &H30B

WM_SPOOLERSTATUS
Global Const WM_SPOOLERSTATUS = &H2A

WM_SYSCHAR
Global Const WM_SYSCHAR = &H106

WM_SYSCOLORCHANGE
Global Const WM_SYSCOLORCHANGE = &H15

WM_SYSCOMMAND
Global Const WM_SYSCOMMAND = &H112

WM_SYSDEADCHAR
Global Const WM_SYSDEADCHAR = &H107

WM_SYSKEYDOWN
Global Const WM_SYSKEYDOWN = &H104

WM_SYSKEYUP
Global Const WM_SYSKEYUP = &H105

WM_TIMECHANGE
Global Const WM_TIMECHANGE = &H1E

WM_TIMER
Global Const WM_TIMER = &H113

WM_UNDO
Global Const WM_UNDO = &H304

WM_USER
Global Const WM_USER = &H400

WM_VKEYTOITEM
Global Const WM_VKEYTOITEM = &H2E

WM_VSCROLL
Global Const WM_VSCROLL = &H115

WM_VSCROLLCLIPBOARD
Global Const WM_VSCROLLCLIPBOARD = &H30A

WM_WINDOWPOSCHANGED
Global Const WM_WINDOWPOSCHANGED = &H0047

WM_WINDOWPOSCHANGING
Global Const WM_WINDOWPOSCHANGING = &H0046

WM_WININICHANGE
Global Const WM_WININICHANGE = &H1A

WN_SUCCESS
Global Const WN_SUCCESS = &H0000

WN_NOT_SUPPORTED
Global Const WN_NOT_SUPPORTED = &H001

WN_NET_ERROR
Global Const WN_NET_ERROR = &H0002

WN_MORE_DATA
Global Const WN_MORE_DATA = &H0003

WN_BAD_POINTER
Global Const WN_BAD_POINTER = &H0004

WN_BAD_VALUE
Global Const WN_BAD_VALUE = &H0005

WN_BAD_PASSWORD
Global Const WN_BAD_PASSWORD = &H0006

WN_ACCESS_DENIED
Global Const WN_ACCESS_DENIED = &H0007

WN_FUNCTION_BUSY
Global Const WN_FUNCTION_BUSY = &H0008

WN_WINDOWS_ERROR
Global Const WN_WINDOWS_ERROR = &H0009

WN_BAD_USER
Global Const WN_BAD_USER = &H000A

WN_OUT_OF_MEMORY
Global Const WN_OUT_OF_MEMORY = &H000B

WN_CANCEL
Global Const WN_CANCEL = &H000C

WN_CONTINUE
Global Const WN_CONTINUE = &H000D

WN_NOT_CONNECTED
Global Const WN_NOT_CONNECTED = &H0030

WN_OPEN_FILES
Global Const WN_OPEN_FILES = &H0031

WN_BAD_NETNAME
Global Const WN_BAD_NETNAME = &H0032

WN_BAD_LOCALNAME
Global Const WN_BAD_LOCALNAME = &H0033

WN_ALREADY_CONNECTED
Global Const WN_ALREADY_CONNECTED = &H0034

WN_DEVICE_ERROR
Global Const WN_DEVICE_ERROR = &H0035

WN_CONNECTION_CLOSED
Global Const WN_CONNECTION_CLOSED = &H0036

WPF_RESTORETOMAXIMIZED
Global Const WPF_RESTORETOMAXIMIZED = &H0002

WPF_SETMINPOSITION
Global Const WPF_SETMINPOSITION = &H0001

WRITEAPI
Global Const WRITEAPI = 1

WS_BORDER
Global Const WS_BORDER = &H800000&

WS_CAPTION
Global Const WS_CAPTION = &HC00000&

WS_CHILD
Global Const WS_CHILD = &H40000000&

WS_CHILDWINDOW
Global Const WS_CHILDWINDOW = (WS_CHILD)

WS_CLIPCHILDREN
Global Const WS_CLIPCHILDREN = &H2000000&

WS_CLIPSIBLINGS
Global Const WS_CLIPSIBLINGS = &H4000000&

WS_DISABLED
Global Const WS_DISABLED = &H8000000&

WS_DLGFRAME
Global Const WS_DLGFRAME = &H400000&

WS_EX_ACCEPTFILES
Global Const WS_EX_ACCEPTFILES = &H00000010&

WS_EX_DLGMODALFRAME
Global Const WS_EX_DLGMODALFRAME = &H00001&

WS_EX_NOPARENTNOTIFY
Global Const WS_EX_NOPARENTNOTIFY = &H00004&

WS_EX_TOPMOST
Global Const WS_EX_TOPMOST = &H00000008&

WS_EX_TRANSPARENT
Global Const WS_EX_TRANSPARENT = &H00000020&

WS_GROUP
Global Const WS_GROUP = &H20000&

WS_HSCROLL
Global Const WS_HSCROLL = &H100000&

WS_ICONIC
Global Const WS_ICONIC = WS_MINIMIZE

WS_MAXIMIZE
Global Const WS_MAXIMIZE = &H1000000&

WS_MAXIMIZEBOX
Global Const WS_MAXIMIZEBOX = &H10000&

WS_MINIMIZE
Global Const WS_MINIMIZE = &H20000000&

WS_MINIMIZEBOX
Global Const WS_MINIMIZEBOX = &H20000&

WS_OVERLAPPED
Global Const WS_OVERLAPPED = &H00000&

WS_OVERLAPPEDWINDOW
Global Const WS_OVERLAPPEDWINDOW = (WS_OVERLAPPED Or WS_CAPTION Or WS_SYSMENU Or WS_THICKFRAME Or WS_MINIMIZEBOX Or WS_MAXIMIZEBOX)

WS_POPUP
Global Const WS_POPUP = &H80000000&

WS_POPUPWINDOW
Global Const WS_POPUPWINDOW = (WS_POPUP Or WS_BORDER Or WS_SYSMENU)

WS_SIZEBOX
Global Const WS_SIZEBOX = WS_THICKFRAME

WS_SYSMENU

Global Const WS_SYSMENU = &H80000&

WS_TABSTOP

Global Const WS_TABSTOP = &H10000&

WS_THICKFRAME

Global Const WS_THICKFRAME = &H40000&

WS_TILED

Global Const WS_TILED = WS_OVERLAPPED

WS_TILEDWINDOW

Global Const WS_TILEDWINDOW = (WS_OVERLAPPEDWINDOW)

WS_VISIBLE

Global Const WS_VISIBLE = &H10000000&

WS_VSCROLL

Global Const WS_VSCROLL = &H200000&

WVR_ALIGNBOTTOM

Global Const WVR_ALIGNBOTTOM = &H0040

WVR_ALIGNLEFT

Global Const WVR_ALIGNLEFT = &H0020

WVR_ALIGNRIGHT

Global Const WVR_ALIGNRIGHT = &H0080

WVR_ALIGNTOP

Global Const WVR_ALIGNTOP = &H0010

WVR_HREDRAW

Global Const WVR_HREDRAW = &H0100

WVR_REDRAW

Global Const WVR_REDRAW = (WVR_HREDRAW Or WVR_VREDRAW)

WVR_VALIDRECTS

Global Const WVR_VALIDRECTS = &H0400

WVR_VREDRAW

Global Const WVR_VREDRAW = &H0200

Appendix I

API TYPE DECLARATIONS

RECT
```
Type RECT
    left As Integer
    top As Integer
    right As Integer
    bottom As Integer
End Type
```

POINTAPI
```
Type POINTAPI
    x As Integer
    y As Integer
End Type
```

PARAMETERBLOCK
```
Type PARAMETERBLOCK
    wEnvSeg As Integer
    lpCmdLine As Long
    lpCmdShow As Long
    dwReserved As Long
End Type
```

OFSTRUCT
```
Type OFSTRUCT
    cBytes As String * 1
    fFixedDisk As String * 1
    nErrCode As Integer
    reserved As String * 4
    szPathName As String * 128
End Type
```

BITMAP
```
Type BITMAP
    bmType As Integer
    bmWidth As Integer
    bmHeight As Integer
    bmWidthBytes As Integer
    bmPlanes As String * 1
    bmBitsPixel As String * 1
    bmBits As Long
End Type
```

RGBTRIPLE
```
Type RGBTRIPLE
    rgbtBlue As String * 1
    rgbtGreen As String * 1
    rgbtRed As String * 1
End Type
```

RGBQUAD
```
Type RGBQUAD
    rgbBlue as String * 1
    rgbGreen As String * 1
    rgbRed As String * 1
    rgbReserved As String * 1
End Type
```

BITMAPCOREHEADER
```
Type BITMAPCOREHEADER
    bcSize as Long
    bcWidth As Integer
    bcHeight As Integer
    bcPlanes As Integer
    bcBitCount As Integer
End Type
```

BITMAPINFOHEADER
```
Type BITMAPINFOHEADER
    biSize As Long
    biWidth As Long
    biHeight As Long
    biPlanes As Integer
    biBitCount As Integer
    biCompression As Long
    biSizeImage As Long
    biXPelsPerMeter As Long
    biYPelsPerMeter As Long
    biClrUsed As Long
    biClrImportant As Long
End Type
```

BITMAPINFO
```
Type BITMAPINFO
    bmiHeader as BITMAPINFOHEADER
    bmiColors As String * 128
End Type
```

BITMAPCOREINFO
```
Type BITMAPCOREINFO
    bmciHeader As BITMAPCOREHEADER
    bmciColors As String * 96
End Type
```

BITMAPFILEHEADER
```
Type BITMAPFILEHEADER
    bfType As Integer
    bfSize As Long
    bfReserved1 As Integer
    bfReserved2 As Integer
    bfOffBits As Long
End Type
```

HANDLETABLE
```
Type HANDLETABLE
    objectHandle As String * 512
End Type
```

METARECORD
```
Type METARECORD
    rdSize As Long
    rdFunction As Integer
    rdParm As String * 512
End Type
```

METAFILEPICT
```
Type METAFILEPICT
    mm As Integer
    xExt As Integer
    yExt As Integer
    hMF As Integer
End Type
```

METAHEADER
```
Type METAHEADER
    mtType As Integer
    mtHeaderSize As Integer
    mtVersion As Integer
    mtSize As Long
    mtNoObjects As Integer
    mtMaxRecord As Long
    mtNoParameters As Integer
End Type
```

TEXTMETRIC
```
Type TEXTMETRIC
    tmHeight As Integer
    tmAscent As Integer
    tmDescent As Integer
    tmInternalLeading As Integer
    tmExternalLeading As Integer
    tmAveCharWidth As Integer
    tmMaxCharWidth As Integer
    tmWeight As Integer
    tmItalic As String * 1
    tmUnderlined As String * 1
    tmStruckOut As String * 1
    tmFirstChar As String * 1
    tmLastChar As String * 1
    tmDefaultChar As String * 1
    tmBreakChar As String * 1
    tmPitchAndFamily As String * 1
    tmCharSet As String * 1
    tmOverhang As Integer
    tmDigitizedAspectX As Integer
    tmDigitizedAspectY As Integer
End Type
```

PELARRAY
```
Type PELARRAY
    paXCount As Integer
    paYCount As Integer
    paXExt As Integer
    paYExt As Integer
    paRGBs As Integer
End Type
```

LOGBRUSH
```
Type LOGBRUSH
    lbStyle As Integer
    lbColor As Long
    lbHatch As Integer
End Type
```

LOGPEN
```
Type LOGPEN
    lopnStyle As Integer
    lopnWidth As POINTAPI
    lopnColor As Long
End Type
```

PALETTEENTRY
```
Type PALETTEENTRY
    peRed As String * 1
    peGreen As String * 1
    peBlue As String * 1
    peFlags As String * 1
End Type
```

LOGPALETTE
```
Type LOGPALETTE
    palVersion As Integer
    palNumEntries As Integer
    palPalEntry As String * 250
End Type
```

LOGFONT

```
Type LOGFONT
    lfHeight As Integer
    lfWidth As Integer
    lfEscapement As Integer
    lfOrientation As Integer
    lfWeight As Integer
    lfItalic As String * 1
    lfUnderline As String * 1
    lfStrikeOut As String * 1
    lfCharSet As String * 1
    lfOutPrecision As String * 1
    lfClipPrecision As String * 1
    lfQuality As String * 1
    lfPitchAndFamily As String * 1
    lfFaceName As String * LF_FACESIZE
End Type
```

EVENTMSG

```
Type EVENTMSG
    message As Integer
    paramL As Integer
    paramH As Integer
    time As Long
End Type
```

MSG

```
Type MSG
    hwnd As Integer
    message As Integer
    wParam As Integer
    lParam As Long
    time As Long
    pt As POINTAPI
End Type
```

PAINTSTRUCT

```
Type PAINTSTRUCT
    hdc As Integer
    fErase As Integer
    rcPaint As RECT
    fRestore As Integer
    fIncUpdate As Integer
    rgbReserved As String * 16
End Type
```

CREATESTRUCT

```
Type CREATESTRUCT
    lpCreateParams As Long
    hInstance As Integer
    hMenu As Integer
    hwndParent As Integer
    cy As Integer
    cx As Integer
    y As Integer
    x As Integer
    style As Long
    lpszName As Long
    lpszClass As Long
    ExStyle As Long
End Type
```

MEASUREITEMSTRUCT

```
Type MEASUREITEMSTRUCT
    CtlType As Integer
    CtlID As Integer
    itemID As Integer
    itemWidth As Integer
    itemHeight As Integer
    itemData As Long
End Type
```

DRAWITEMSTRUCT
```
Type DRAWITEMSTRUCT
    CtlType As Integer
    CtlID As Integer
    itemID As Integer
    itemAction As Integer
    itemState As Integer
    hwndItem As Integer
    hDC As Integer
    rcItem As RECT
    itemData As Long
End Type
```

DELETEITEMSTRUCT
```
Type DELETEITEMSTRUCT
    CtlType As Integer
    CtlID As Integer
    itemID As Integer
    hwndItem As Integer
    itemData As Long
End Type
```

COMPAREITEMSTRUCT
```
Type COMPAREITEMSTRUCT
    CtlType As Integer
    CtlID As Integer
    hwndItem As Integer
    itemID1 As Integer
    itemData1 As Long
    itemID2 As Integer
    itemData2 As Long
End Type
```

MENUITEMTEMPLATEHEADER
```
Type MENUITEMTEMPLATEHEADER
    versionNumber As Integer
    offset As Integer
End Type
```

MENUITEMTEMPLATE
```
Type MENUITEMTEMPLATE
    mtOption As Integer
    mtID As Integer
    mtString As Long
End Type
```

DCB
```
Type DCB
    Id As String * 1
    BaudRate As Integer
    ByteSize As String * 1
    Parity As String * 1
    StopBits As String * 1
    RlsTimeout As Integer
    CtsTimeout As Integer
    DsrTimeout As Integer
    Bits1 As String * 1
    Bits2 As String * 1
    XonChar As String * 1
    XoffChar As String * 1
    XonLim As Integer
    XoffLim As Integer
    PeChar As String * 1
    EofChar As String * 1
    EvtChar As String * 1
    TxDelay As Integer
End Type
```

COMSTAT
```
Type COMSTAT
    Bits As String * 1
    cbInQue As Integer
    cbOutQue As Integer
End Type
```

MDICREATESTRUCT
```
Type MDICREATESTRUCT
    szClass As Long
    szTitle As Long
    hOwner As Integer
    x As Integer
    y As Integer
    cx As Integer
    cy As Integer
    style As Long
    lParam As Long
End Type
```

CLIENTCREATESTRUCT
```
Type CLIENTCREATESTRUCT
    hWindowMenu As Integer
    idFirstChild As Integer
End Type
```

MULTIKEYHELP
```
Type MULTIKEYHELP
    mkSize As Integer
    mkKeylist As String * 1
    szKeyphrase As String * 250
End Type
```

WINDEBUGINFO
```
Type WINDEBUGINFO
    flags As Integer
    dwOptions As Long
    dwFilter As Long
    achAllocModule As String * 8
    dwAllocBreak As Long
    dwAllocCount As Long
End Type
```

SIZE
```
Type SIZE
    cx As Integer
    cy As Integer
End Type
```

PANOSE
```
Type PANOSE
    bFamilyType As String * 1
    bSerifStyle As String * 1
    bWeight As String * 1
    bProportion As String * 1
    bContrast As String * 1
    bStrokeVariation As String * 1
    bArmStyle As String * 1
    bLetterform As String * 1
    bMidline As String * 1
    bXHeight As String * 1
End Type
```

OUTLINETEXTMETRIC
```
Type OUTLINETEXTMETRIC
    otmSize As Integer
    otmTextMetrics As TEXTMETRIC
    otmFiller As String * 1
    otmPanoseNumber As PANOSE
    otmfsSelection As Integer
    otmfsType As Integer
    otmsCharSlopeRise As Integer
    otmsCharSlopeRun As Integer
    otmItalicAngle As Integer
    otmEMSquare As Integer
    otmAscent As Integer
    otmDescent As Integer
    otmLineGap As Integer
    otmsCapEmHeight As Integer
    otmsXHeight As Integer
    otmrcFontBox As RECT
    otmMacAscent As Integer
    otmMacDescent As Integer
    otmMacLineGap As Integer
    otmusMinimumPPEM As Integer
    otmptSubscriptSize As POINTAPI
    otmptSubscriptOffset As POINTAPI
    otmptSuperscriptSize As POINTAPI
    otmptSuperscriptOffset As POINTAPI
    otmsStrikeoutSize As Integer
    otmsStrikeoutPosition As Integer
    otmsUnderscorePosition As Integer
    otmsUnderscoreSize As Integer
    otmpFamilyName As Long
    otmpFaceName As Long
    otmpStyleName As Long
    otmpFullName As Long
End Type
```

NEWTEXTMETRIC
```
Type NEWTEXTMETRIC
    tmHeight As Integer
    tmAscent As Integer
    tmDescent As Integer
    tmInternalLeading As Integer
    tmExternalLeading As Integer
    tmAveCharWidth As Integer
    tmMaxCharWidth As Integer
    tmWeight As Integer
    tmItalic As String * 1
    tmUnderlined As String * 1
    tmStruckOut As String * 1
    tmFirstChar As String * 1
    tmLastChar As String * 1
    tmDefaultChar As String * 1
    tmBreakChar As String * 1
    tmPitchAndFamily As String * 1
    tmCharSet As String * 1
    tmOverhang As Integer
    tmDigitizedAspectX As Integer
    tmDigitizedAspectY As Integer
    ntmFlags As Long
    ntmSizeEM As Integer
    ntmCellHeight As Integer
    ntmAvgWidth As Integer
End Type
```

GLYPHMETRICS
```
Type GLYPHMETRICS
    gmBlackBoxX As Integer
    gmBlackBoxY As Integer
    gmptGlyphOrigin As POINTAPI
    gmCellIncX As Integer
    gmCellIncY As Integer
End Type
```

FIXED
```
Type FIXED
    fract As Integer
    value As Integer
End Type
```

MAT2
```
Type MAT2
    eM11 As FIXED
    eM12 As FIXED
    eM21 As FIXED
    eM22 As FIXED
End Type
```

POINTFX
```
Type POINTFX
    x As FIXED
    y As FIXED
End Type
```

TTPOLYCURVE
```
Type TTPOLYCURVE
    wType As Integer
    cpfx As Integer
    apfx As POINTFX
End Type
```

TTPOLYGONHEADER
```
Type TTPOLYGONHEADER
    cb As Long
    dwType As Long
    pfxStart As POINTFX
End Type
```

ABC
```
Type ABC
    abcA As Integer
    abcB As Integer
    abcC As Integer
End Type
```

KERNINGPAIR
```
Type KERNINGPAIR
    wFirst As Integer
    wSecond As Integer
    iKernAmount As Integer
End Type
```

RASTERIZER_STATUS
```
Type RASTERIZER_STATUS
    nSize As Integer
    wFlags As Integer
    nLanguageID As Integer
End Type
```

DOCINFO
```
Type DOCINFO
    cbSize As Integer
    lpszDocName As Long
    lpszOutput As Long
End Type
```

WINDOWPLACEMENT
```
Type WINDOWPLACEMENT
    length As Integer
    flags As Integer
    showCmd As Integer
    ptMinPosition As POINTAPI
    ptMaxPosition As POINTAPI
    rcNormalPosition As RECT
End Type
```

WINDOWPOS
```
Type WINDOWPOS
    hwnd As Integer
    hwndInsertAfter As Integer
    x As Integer
    y As Integer
    cx As Integer
    cy As Integer
    flags As Integer
End Type
```

NCCALCSIZE_PARAMS
```
Type NCCALCSIZE_PARAMS
    rgrc As Long
    lppos As Long
End Type
```

CBT_CREATEWND
```
Type CBT_CREATEWND
    lpcs As Long
    hwndInsertAfter As Integer
End Type
```

CBTACTIVATESTRUCT
```
Type CBTACTIVATESTRUCT
    fMouse As Integer
    hWndActive As Integer
End Type
```

HARDWAREHOOKSTRUCT
```
Type HARDWAREHOOKSTRUCT
    hWnd As Integer
    wMessage As Integer
    wParam As Integer
    lParam As Long
End Type
```

DEBUGHOOKINFO
```
Type DEBUGHOOKINFO
    hModuleHook As Integer
    reserved As Long
    lParam As Long
    wParam As Integer
    code As Integer
End Type
```

DRVCONFIGINFO
```
Type DRVCONFIGINFO
    dwDCISize As Long
    lpszDCISectionName As Long
    lpszDCIAliasName As Long
End Type
```

DRIVERINFOSTRUCT
```
Type DRIVERINFOSTRUCT
    length As Integer
    hDriver As Integer
    hModule As Integer
    szAliasName As String * 128
End Type
```

Appendix J

COMMON PROPERTIES, METHODS AND EVENTS

This appendix provides a quick reference to the properties, methods and event of some of the VB components. This is not meant to be a complete reference but is fast and easy to review. For more in depth information, you should consult Microsoft's Visual Basic Language Reference and ActiveX Controls Reference are highly recommended.

There are several properties such as BackColor that are common to many controls. This first section will detail these common properties.

Common Properties

Appearance	Integer	FontUnderline	Boolean
BackColor	Long	ForeColor	Long
BorderStyle	Integer	hDC	Long
Container	Object	Height	Single
DataBindings	DataBindings Collection	HelpContextID	Long
		hWnd	Long
DataChanged	Boolean	Index	Integer
DataField	String	Left	Single
DragIcon	IPictureDisp Object	LinkItem	String
DragMode	Integer	LinkMode	Integer
Enabled	Boolean	LinkTopic	String
Font	StdFont Object or IFontDisp Object	LinkTimeout	Integer
		MouseIcon	IPictureDisp Object
FontBold	Boolean	MousePointer	Integer
FontItalic	Boolean	Name	String
FontName	String	Object	Object
FontSize	Single	OLEDragMode	Integer
FontStrikethru	Boolean	OLEDropMode	Integer
FontTransparent	Boolean	Parent	Form Object or Object

399

Common Properties (continued)

RightToLeft	Boolean	TabStop	Boolean
ScaleHeight	Single	Tag	String
ScaleLeft	Single	ToolTipText	String
ScaleMode	Integer	Top	Single
ScaleTop	Single	Visible	Boolean
ScaleWidth	Single	WhatsThisHelpID	Long
TabIndex	Integer	Width	Single

Common Methods

Drag	object.Drag(action)	Scale	object.Scale(x1, y1)-(x2, y2)
LinkExecute	object.LinkExecute (cmdstr, cancel)	ScaleX	object.ScaleX(width, fromscale, toscale)
LinkPoke	object.LinkPoke	ScaleY	object.ScaleY(height, fromscale, toscale)
LinkRequest	object.LinkRequest	SetFocus	object.SetFocus
LinkSend	object.LinkSend	ShowWhatsThis	object.ShowWhatsThis
Move	object.Move (rows, start)	Zorder	object.ZOrder(position)
OLEDrag	object.OLEDrag		
Refresh	object.Refresh		

Common Events

Click()

DblClick()

DragDrop(source As Control, x As Single, y As Single)

DragOver(source As Control, x As Single, y As Single, state As Integer)

GotFocus()

KeyDown(keycode As Integer, shift As Integer)

KeyPress(keyascii As Integer)

KeyUp(keycode As Integer, shift As Integer)

LinkClose()

LinkError(linkerr As Integer)

LinkExecute(cmdstr As String, cancel As Integer)

LinkNotify([index As Integer])

LinkOpen(cancel As Integer)

LostFocus()

MouseDown(button As Integer, shift As Integer, x As Single, y As Single)

MouseMove(button As Integer, shift As Integer, x As Single, y As Single)

MouseUp(button As Integer, shift As Integer, x As Single, y As Single)

OLECompleteDrag(effect As Long)

Common Events (continued)

OLEDragDrop(data As DataObject, effect As Long, button As Integer, shift As Integer, x As Single, y As Single)

OLEDragOver(data As DataObject, effect As Long, button As Integer, shift As Integer, x As Single, y As Single, state As Integer)

OLEGiveFeedback(effect As Long, defaultcursors As Boolean)

OLESetData(data As DataObject, dataFormat As Integer)

OLEStartDrag(data As DataObject, allowedeffects As Long)

SPECIFICS FOR INDIVIDUAL OBJECTS AND CONTROLS

Animation
Specific Properties
AutoPlay	Boolean
BackStyle	Integer
Center	Boolean

Specific Methods
Close	object.Close
Open	object.Open (file)
Play	object.Play ([repeatcount], [startframe], [endframe])
Stop	object.Stop

CheckBox
Specific Properties
Alignment	Integer
Caption	String
DisabledPicture	IPictureDisp Object
DownPicture	IPictureDisp Object
MaskColor	Long
Picture	IPictureDisp Object
Style	Integer
UseMaskColor	Boolean
Value	Integer

ComboBox
Specific Properties
IntegralHeight	Boolean
ItemData	Long Array
List	String Array
ListCount	Integer
ListIndex	Integer
Locked	Boolean
NewIndex	Integer
SelLength	Long
SelStart	Long
SelText	String
Sorted	Boolean
Style	Integer
Text	String
TopIndex	Integer

Specific Methods
AddItem	object.AddItem(item [, index])
Clear	object.Clear
RemoveItem	object.RemoveItem (index)

CommandButton
Specific Properties
Cancel	Boolean
Caption	String
Default	Boolean
DisabledPicture	IPictureDisp Object
DownPicture	IPictureDisp Object
MaskColor	Long
Picture	IPictureDisp Object
Style	Integer
UseMaskColor	Boolean
Value	Boolean

Common Dialog
Properties
Action	Integer
CancelError	Boolean
Color	OLE_COLOR
Copies	Integer
DefaultExt	String
DialogTitle	String
FileName	String
FileTitle	String
Filter	String
FilterIndex	Integer
Flags	Long
HelpCommand	Integer
HelpContext	Long
HelpFile	String
HelpKey	String
InitDir	String
Max	Integer
MaxFileSize	Integer
Min	Integer
PrinterDefault	Boolean
ToPage	Integer

Specific Methods
ShowColor	object.ShowColor
ShowFont	object.ShowFont
ShowHelp	object.ShowHelp
ShowOpen	object.ShowOpen
ShowPrinter	object.ShowPrinter
ShowSave	object.ShowSave

Data
Specific Properties
Align	Integer
BOFAction	Integer
Caption	String
Connect	String
Database	Database Object
DatabaseName	String
DefaultCursorType	Integer
DefaultType	Integer
EditMode	Integer
EOFAction	Integer
Exclusive	Boolean
Options	Integer
ReadOnly	Boolean
Recordset	Recordset Object
RecordsetType	Integer
RecordSource	String

Specific Methods
UpdateControls	object.UpdateControls
UpdateRecord	object.UpdateRecord

DBCombo
Specific Properties
BoundColumn	String
BoundText	String
IntegralHeight	Boolean
ListField	String
Locked	Boolean
MatchedWithList	Boolean
MatchEntry	MatchEntryConstants Enum

RowSource	IRowCursor Object	hWndEditor	OLE_HANDLE Object
SelectedItem	Variant	LeftCol	Integer
SelLength	Long	MarqueeStyle	enumMarquee-StyleConstants Enum
SelStart	Long		
SelText	String		
Style	StyleConstants Enum		
VisibleCount	Integer	MarqueeUnique	Boolean
VisibleItems	Variant Array	RecordSelectors	Boolean
		Row	Integer

Specific Methods

ReFill	object.ReFill

DBGrid

Properties

AddNewMode	enumAddNewModeConstants Enum	RowDividerStyle	enumDividerStyleConstants Enum
		RowHeight	Single
		ScrollBars	enumScrollBarConstants s Enum
Align	Integer		
AllowAddNew	Boolean	SelBookmarks	Variant Array
AllowArrows	Boolean	SelEndCol	Integer
AllowDelete	Boolean	SelLength	Long
AllowRowSizing	Boolean	SelStart	Long
AllowUpdate	Boolean	SelStartCol	Integer
ApproxCount	Long	SelText	String
Bookmark	Variant	Split	Integer
Caption	String	Splits	Object Array
Col	Integer	TabAcrossSplits	Boolean
ColumnHeaders	Boolean	TabAction	enumTabActionConstants Enum
Columns	Object Array		
CurrentCellModified	Boolean	Text	String
CurrentCellVisible	Boolean	VisibleCols	Integer
DataMode	enumDataModeConstants Enum	VisibleRows	Integer
		WrapCellPointer	Boolean

Specific Methods

DataSource	ICursor Object
DefColWidth	Single
EditActive	Boolean
ErrorText	String
FirstRow	Variant
HeadFont	IFontDisp Object
HeadLines	Single

AboutBox	Object.AboutBox
CaptureImage	Object.CaptureImage
ClearFields	Object.ClearFields
ClearSelCols	object.ClearSelCols
ColContaining	object.ColContaining (coordinate)
GetBookmark	object.GetBookmark (value)
HoldFields	object.HoldFields
ReBind	object.Rebind

RowBookmark   object.RowBookmark (value)
RowContaining object.RowContaining (coordinate)
RowTop   object.RowTop (value)
Scroll   object.Scroll (colvalue, rowvalue)

Specific Events
AfterColEdit([index As Integer,] ByVal colindex As Integer)
AfterColUpdate([index As Integer,] colindex As Integer)
AfterDelete([index As Integer,] colindex As Integer)
AfterInsert(index As Integer)
AfterUpdate(index As Integer)
BeforeColEdit([index As Integer,] ByVal colindex As Integer, ByVal keyascii As Integer, cancel As Integer)
BeforeColUpdate([index As Integer,] colindex As Integer, oldvalue As Variant, cancel As Integer)
BeforeDelete([index As Integer,] cancel As Integer)
BeforeInsert([index As Integer,] cancel As Integer)
BeforeUpdate([index As Integer,] cancel As Integer)
ButtonClick([index As Integer,] ByVal colindex As Integer)
Change([index As Integer])
ColEdit([index As Integer,] ByVal colindex As Integer)
ColResize([index As Integer,] colindex As Integer, cancel As Integer)
Error([index As Integer,] ByVal dataerror As Integer, response As Integer)
HeadClick([index As Integer,] colindex As Integer)
OnAddNew([index As Integer])
RowColChange([index As Integer, lastrow As String, lastcol As Integer])
RowResize([index As Integer,] cancel As Integer)
Scroll([cancel As Integer])
SelChange([index As Integer,] cancel As Integer)
SplitChange([index As Integer])
UnboundAddData(rowbuf As RowBuffer, newrowbookmark As Variant)
UnboundDeleteRow(bookmark As Variant)
UnboundGetRelativeBookmark([index As Integer,] startlocation As Variant, ByVal offset As Long, newlocation As Variant, approximateposition As Long)
UnboundReadData(rowbuf As RowBuffer, startlocation As Variant, readpriorrows As Boolean)
UnboundWriteData(rowbuf As RowBuffer, writelocation As Variant)

DBList
Specific Properties
BoundColumn	String
BoundText	String
IntegralHeight	Boolean
ListField	String
Locked	Boolean
MatchedWithList	Boolean
RowSource	IRowCursor Object
SelectedItems	Variant
Text	String
VisibleCount	Integer
VisibleItems	Variant Array

Specific Methods
ReFill	object.ReFill

DirListBox
Properties
List	String Array
ListCount	Integer
ListIndex	Integer
Path	String
TopIndex	Integer

DriveListBox
Properties
Drive	String
List	String Array
ListCount	Integer
ListIndex	Integer
TopIndex	Integer

FileListBox
Properties
Archive	Boolean
FileName	String
Hidden	Boolean
List	String Array
ListCount	Integer
ListIndex	Integer
MultiSelect	Integer
Normal	Boolean
Path	String
Pattern	String
ReadOnly	Boolean
Selected	Boolean Array
System	Boolean
TopIndex	Integer

Form
Specific Properties
ActiveControl	Control Object
AutoRedraw	Boolean
BorderStyle	Integer
Caption	String
ClipControls	Boolean
ControlBox	Boolean
Controls	Object
Count	Integer
CurrentX	Single
CurrentY	Single
DrawMode	Integer
DrawStyle	Integer
DrawWidth	Integer
FillColor	Long
FillStyle	Integer
Icon	IpictureDisp Object
Image	IPictureDisp Object
KeyPreview	Boolean
MaxButton	Boolean
MDIChild	Boolean
MinButton	Boolean
Moveable	Boolean
Palette	IpictureDisp Object
PaletteMode	Integer
Picture	IPictureDisp Object
ShowInTaskbar	Boolean
StartUpPosition	Integer
WhatsThisButton	Boolean
WhatsThisHelp	Boolean
WindowState	Integer

Specific Methods

Circle	object.Circle [Step] (x, y), [color, start, end, aspect]
Cls	object.Cls
Hide	object.Hide
Line	object.Line [Step] (x1, y1) [Step] (x2, y2), [color], [B][F]
PaintPicture	object.PaintPicture(picture, x1, y1, width1, height1, x2, y2, width2, height2, opcode)
Point	object.Point(x, y)
PopupMenu	object.PopupMenu(menuname, flags, x, y, boldcommand)
PrintForm	object.PrintForm
PSet	object.PSet [Step] (x, y), [color]
Show	object.Show(style, ownerform)
TextHeight	object.TextHeight(string)
TextWidth	object.TextWidth(string)
WhatsThisMode	object.WhatsThisMode

Specific Events

Activate()
Deactivate()
Initialize()
Load()
Paint()
QueryUnload(*cancel* As Integer, *unloadmode* As Integer)
Resize()
Terminate()
Unload(*cancel* As Integer)

Frame

Properties

Caption	String
ClipControls	Boolean

Horizontal Scroll Bar
Specific Properties
LargeChange	Integer
Max	Integer
Min	Integer
SmallChange	Integer
Value	Integer

Specific Events
Change([index As Integer])
Scroll()

Image
Properties
DataChanged	Boolean
DataField	String
Picture	IPictureDisp Object
Stretch	Boolean

ImageList
Specific Properties
hImageList	OLE_HANDLE Object
ImageHeight	Integer
ImageWidth	Integer
ListImages	ListImages Collection
MaskColor	OLE_COLOR Object
UseMaskColor	Boolean

Specific Methods
Overlay	object.Overlay(key1, key2) As IpictureDisp

Inet
Properties
AccessType	AccessConstants Enum
Document	String
hInternet	Long
Password	String
Protocol	ProtocolConstants Enum
Proxy	String
RemoteHost	String
RemotePort	Integer
RequestTimeout	Long
ResponseCode	Long
ResponseInfo	String
StillExecuting	Boolean
URL	String
UserName	String

Specific Methods
Cancel	object.Cancel
Execute	object.Execute(url, operation, data, requestheaders)
GetChunk	object.GetChunk(size [, datatype])
GetHeader	object.GetHeader(hrdname)
OpenURL	object.OpenURL(url [, datatype])

Specific Events
StateChanged(ByVal *state* As Integer)

Label
Properties
Alignment	Integer
AutoSize	Boolean
BackStyle	Integer
Caption	String
UseMnemonic	Boolean
WordWrap	Boolean

Specific Events
Change()

Line
Properties
BorderColor	Long
BorderWidth	Integer
DrawMode	Integer
X1	Single
X2	Single
Y1	Single
Y2	Single

ListBox
Properties
Columns	Integer
IntegralHeight	Boolean
ItemData	Long Array
List	String Array
ListCount	Integer
ListIndex	Integer
MultiSelect	Integer
NewIndex	Integer
SelCount	Integer
Selected	Boolean Array
Sorted	Boolean
Style	Integer
Text	String
TopIndex	Integer

Methods
AddItem	object.AddItem(item [, index])
Clear	object.Clear
RemoveItem	object.RemoveItem (index)

Events
ItemCheck(item As Integer)
Scroll()

ListView
Properties
ColumnHeaders	ColumnHeaders Collection
DropHighlight	ListItem Object
HideColumnHeader	Boolean
HideSelection	Boolean
Icons	Object
LabelEdit	ListLabelEditConstants
LabelWrap	Boolean
ListItems	ListItems Collection
MultiSelect	Boolean
SelectedItem	ListItem Object
SmallIcons	Object
Sorted	Boolean
SortKey	Integer
SortOrder	ListSortOrderConstants
View	ListViewConstants

Methods
FindItem	object.FindItem (string, [value], [index], [match]) As ListItem
GetFirstVisible	object.GetFirstVisible() As ListItem
HitTest	object.HitTest (x, y) As ListItem
StartLabelEdit	object.StartLabelEdit

Events
AfterLabelEdit(cancel As Integer, newstring As String)
BeforeLabelEdit(cancel As Integer)
ColumnClick(colheader As ColumnHeader)
ItemClick(item As ListItem)

MAPI Message

Properties

Action	Integer
AddressCaption	String
AddressLabel	String
AddressModifiable	Boolean
AddressResolveUI	Boolean
AttachmentCount	Long
AttachmentIndex	Long
AttachmentName	String
AttachmentPathName	String
AttachmentPosition	Long
AttachmentType	Integer
FetchMsgType	String
FetchSorted	Boolean
FetchUnreadOnly	Boolean
MsgConversationID	String
MsgCount	Long
MsgDateReceived	String
MsgID	String
MsgIndex	Long
MsgNoteText	String
MsgOrigAddress	String
MsgOrigDisplayName	String
MsgRead	Boolean
MsgReceiptRequested	Boolean
MsgSent	Boolean
MsgSubject	String
MsgType	String
RecipAddress	String
RecipCount	Long
RecipDisplayName	String
RecipIndex	Long
RecipType	Integer
SessionID	Long

Methods

Compose	object.Compose
Copy	object.Copy
Delete	object.Delete ([value])
Fetch	object.Fetch
Forward	object.Forward
Reply	object.Reply
ReplyAll	object.ReplyAll
ResolveName	object.ResolveName
Save	object.Save
Send	object.Send ([value])
Show	object.Show ([value])

MAPI Session

Properties

Action	Integer
DownloadMail	Boolean
LogonUI	Boolean
NewSession	Boolean
Password	String
SessionID	Long
UserName	String

Methods

SignOff	object.SignOff
SignOn	object.SignOn

Masked Edit

Properties

AllowPrompt	Boolean
AutoTab	Boolean
ClipMode	ClipModeConstants
ClipText	String
Format	String
FormattedText	String
Mask	String
MaxLength	Integer
PromptChar	String
PromptInclude	Boolean
SelLength	Long
SelStart	Long
SelText	Long
Text	String

Events

Change()
ValidationError(invalidtext As String, startpos As Integer)

MDIForm

Properties
ActiveControl	Control Object
ActiveForm	Object
AutoShowChildren	Boolean
Caption	String
Controls	Object Collection
Count	Integer
Icon	IPictureDisp Object
Moveable	Boolean
Picture	IPictureDisp Object
ScrollBars	Boolean
StartUpPosition	Integer
WhatsThisHelp	Boolean
WindowState	Integer

Methods
Arrange	object.Arrange (arrangement)
Hide	object.Hide
PopupMenu	object.PopupMenu (menu, [flags], [x], [y], [defaultmenu])
Show	object.Show([modal], [ownerform])
WhatsThisMode	object.WhatsThisMode

Events
Activate()
Deactivate()
Initialize()
Load()
QueryUnload(cancel As Integer, unloadmode As Integer)
Resize()
Terminate()
Unload(cancel As Integer)

Microsoft Tabbed Dialog

Properties
Caption	String
Rows	Integer
ShowFocusRect	Boolean
Style	StyleConstants
Tab	Integer
TabCaption	String Array
TabEnabled	Boolean Array
TabHeight	Single
TabMaxWidth	Single
TabOrientation	TabOrientationConstants
TabPicture	IPictureDisp Array
Tabs	Integer
TabsPerRow	Integer
TabVisible	Boolean Array
WordWrap	Boolean

Microsoft Chart

Properties
ActiveSeriesCount	Integer
AllowDithering	Boolean
AllowSelections	Boolean
AllowSeriesSelection	Boolean
AutoIncrement	Boolean
BackDrop	Backdrop Object
BorderStyle	VtBorderStyle Enum
Chart3d	Boolean
ChartData	Variant
ChartType	VtChChartType Enum
Column	Integer
ColumnCount	Integer
ColumnLabel	String
ColumnLabelCount	Integer
ColumnLabelIndex	Integer
Data	String

DataGrid	DataGrid Object	RowLabel	String
DoSetCursor	Boolean	RowLabelCount	Integer
DrawMode	VtChDrawMode Enum	RowLabelIndex	Integer
		SeriesColumn	Integer
Footnote	Footnote Object	SeriesType	VtChSeriesType Enum
FootnoteText	String		
Legend	Legend Object	ShowLegend	Boolean
Plot	Plot Object	Stacking	Boolean
RandomFill	Boolean	TextLengthType	VtTextLengthType Enum
Repaint	Boolean		
Row	Integer	Title	Title Object
RowCount	Integer	TitleText	String

Methods

AboutBox	object.AboutBox
EditCopy	object.EditCopy
EditPaste	object.EditPaste
GetSelectedPart	object.GetSelectedPart (part, index1, index2, index3, index4)
Layout	object.Layout
SelectPart	object.SelectPart (part, index1, index2, index3, index4)
ToDefaults	object.ToDefaults
TwipsToChartPart	object.TwipsToChartPart (xval, yval, part, index1, index2, index3, index4)

Events

AxisActivated(axisid As Integer, axisindex As Integer, mouseflag As Integer, cancel As Integer)

AxisLabelActivated(axisid As Integer, axisindex As Integer, labelsetindex As Integer, labelindex As Integer, mouseflag As Integer, cancel As Integer)

AxisLabelSelected(axisid As Integer, axisindex As Integer, labelsetindex As Integer, labelindex As Integer, mouseflag As Integer, cancel As Integer)

AxisLabelUpdated(axisid As Integer, axisindex As Integer, labelsetindex As Integer, labelindex As Integer, updateflags As Integer)

AxisSelected(axisid As Integer, axisindex As Integer, mouseflag As Integer, cancel As Integer)

AxisTitleActivated(axisid As Integer, axisindex As Integer, mouseflag As Integer, cancel As Integer)

AxisTitleSelected(axisid As Integer, axisindex As Integer, mouseflag As Integer, cancel As Integer)

AxisTitleUpdated(axisid As Integer, axisindex As Integer, updateflags As Integer)
AxisUpdated(axisid As Integer, axisindex As Integer, updateflags As Integer)
ChartActivated(mouseflag As Integer, cancel As Integer)
ChartSelected(mouseflag As Integer, cancel As Integer)
ChartUpdated(updateflags As Integer)
DataUpdated(row As Integer, column As Integer, labelrow As Integer, labelcolumn As Integer, labelsetindex As Integer, updateflags As Integer)
DonePainting()
FootnoteActivated(mouseflag As Integer, cancel As Integer)
FootnoteSelected(mouseflag As Integer, cancel As Integer)
FootnoteUpdated(updateflags As Integer)
LegendActivated(mouseflag As Integer, cancel As Integer)
LegendSelected(mouseflag As Integer, cancel As Integer)
LegendUpdated(updateflags As Integer)
PlotActivated(mouseflag As Integer, cancel As Integer)
PlotSelected(mouseflag As Integer, cancel As Integer)
PlotUpdated(updateflags As Integer)
PointActivated(series As Integer, datapoint As Integer, mouseflag As Integer, cancel As Integer)
PointLabelActivated(series As Integer, datapoint As Integer, mouseflag As Integer, cancel As Integer)
PointLabelSelected(series As Integer, datapoint As Integer, mouseflag As Integer, cancel As Integer)
PointLabelUpdated(series As Integer, datapoint As Integer, updateflags As Integer)
PointSelected(series As Integer, datapoint As Integer, mouseflag As Integer, cancel As Integer)
PointUpdated(series As Integer, datapoint As Integer, updateflags As Integer)
SeriesActivated(series As Integer, mouseflag As Integer, cancel As Integer)
SeriesSelected(series As Integer, mouseflag As Integer, cancel As Integer)
SeriesUpdated(series As Integer, updateflags As Integer)
TitleActivated(mouseflag As Integer, cancel As Integer)
TitleSelected(mouseflag As Integer, cancel As Integer)
TitleUpdated(updateflags As Integer)

MSComm
Properties
Break	Boolean
CDHolding	Boolean
CommEvent	Integer
CommID	Long
CommPort	Integer
CTSHolding	Boolean
DSRHolding	Boolean
DTREnable	Boolean
EOFEnable	Boolean
Handshaking	HandshakeConstants Enum
InBufferCount	Integer
InBufferSize	Integer
Input	Variant
InputLen	Integer
InputMode	InputModeConstants Enum
NullDiscard	Boolean
OutBufferCount	Integer
OutBufferSize	Integer
Output	Variant
ParityReplace	String
PortOpen	Boolean
RThreshold	Integer
RTSEnable	Boolean
Settings	String
SThreshold	Integer

Events
OnComm()

Microsoft Flex Grid
Properties
AllowBigSelection	Boolean
AllowUserResizing	AllowUserResizingSettings Enum
BackColorBkg	OLE_COLOR Object
BackColorFixed	OLE_COLOR Object
BackColorSel	OLE_COLOR Object
CellAlignment	Integer
CellBackColor	OLE_COLOR Object
CellFontBold	Boolean
CellFontItalic	Boolean
CellFontName	String
CellFontSize	Single
CellFontUnderline	Boolean
CellFontWidth	Single
CellForeColor	OLE_COLOR Object
CellHeight	Long
CellLeft	Long
CellPicture	IPictureDisp Object
CellPictureAlignment	Integer
CellTextStyle	TextStyleSettings Enum
CellTop	Long

CellWidth	Long
Clip	String
Col	Long
ColAlignment	Integer Array
ColData	Long Array
ColIsVisible	Boolean Array
ColPos	Long Array
ColPosition	Long Array
Cols	Long
ColSel	Long
ColWidth	Long Array
DataSource	IRowCursor Object
FillStyle	FillStyleSettings Enum
FixedAlignment	Integer Array
FixedCols	Long
FixedRows	Long
FocusRect	FocusRectSettings Enum
FontWidth	Single
ForeColorFixed	OLE_COLOR Object
ForeColorSel	OLE_COLOR Object
FormatString	String
GridColor	OLE_COLOR Object
GridColorFixed	OLE_COLOR Object
GridLines	GridLineSettings Enum
GridLinesFixed	GridLineSettings Enum
GridLineWidth	Integer
HighLight	HighLightSettings Enum
LeftCol	Long
MergeCells	MergeCellsSettings Enum
MergeCol	Boolean Array
MergeRow	Boolean Array
MouseCol	Long
MouseRow	Long
Picture	IpictureDisp Object
PictureType	PictureTypeSettings Enum
Redraw	Boolean
Row	Long
RowData	Long Array
RowHeight	Long Array
RowHeightMin	Long

RowIsVisible	Boolean Array
RowPos	Long Array
RowPosition	Long Array
Rows	Long
RowSel	Long
ScrollBars	ScrollBarsSettings Enum
ScrollTrack	Boolean
SelectionMode	SelectionModeSettings Enum
Sort	Integer
Text	String
TextArray	String Array
TextMatrix	String Array
TextStyle	TextStyleSettings Enum
TextStyleFixed	TextStyleSettings Enum
TopRow	Long
Version	Integer
WordWrap	Boolean

Methods

AddItem	object.AddItem (item, [index])
Clear	object.Clear
RemoveItem	object.RemoveItem (index)

Events

Compare(row1 As Integer, row2 As Integer, cmp As Integer)
EnterCell()
LeaveCell()
RowColChange()
Scroll()
SelChange()

Multimedia (MCI) Control

Properties

AutoEnable	Boolean
BackEnabled	Boolean
BackVisible	Boolean
CanEject	Boolean
CanPlay	Boolean
CanRecord	Boolean
CanStep	Boolean
Command	String
DeviceID	Integer
DeviceType	String
EjectEnabled	Boolean
EjectVisible	Boolean
Error	Integer
ErrorMessage	String
FileName	String
Frames	Long
From	Long
hWndDisplay	Long
Length	Long
Mode	Long
NextEnabled	Boolean
NextVisible	Boolean
Notify	Boolean
NotifyMessage	String
NotifyValue	Integer
Orientation	OrientationConstants Enum
PauseEnabled	Boolean
PauseVisible	Boolean
PlayEnabled	Boolean
PlayVisible	Boolean
Position	Long
PrevEnabled	Boolean
PrevVisible	Boolean
RecordEnabled	Boolean
RecordMode	RecordModeConstants Enum
RecordVisible	Boolean
Shareable	Boolean
Silent	Boolean
Start	Long
StepEnabled	Boolean
StepVisible	Boolean
StopEnabled	Boolean
StopVisible	Boolean
TimeFormat	Long
To	Long
Track	Long
TrackLength	Long
TrackPosition	Long
Tracks	Long
UpdateInterval	Integer
UsesWindows	Boolean
Wait	Boolean

Events

BackClick(cancel As Integer)
BackCompleted(errcode As Long)
BackGotFocus()
BackLostFocus()
Done(notifycode As Integer)
EjectClick(cancel As Integer)
EjectCompleted(errcode As Long)
EjectGotFocus()
EjectLostFocus()
NextClick(cancel As Integer)
NextCompleted(errcode As Long)
NextGotFocus()
NextLostFocus()
PauseClick(cancel As Integer)
PauseCompleted(errcode As Long)
PauseGotFocus()
PauseLostFocus()
PlayClick(cancel As Integer)
PlayCompleted(errcode As Long)
PlayGotFocus()
PlayLostFocus()
PrevClick(cancel As Integer)
PrevCompleted(errcode As Long)
PrevGotFocus()
PrevLostFocus()
RecordClick(cancel As Integer)
RecordCompleted(errcode As Long)
RecordGotFocus()
RecordLostFocus()

StatusUpdate()
StepClick(cancel As Integer)
StepCompleted(errcode As Long)
StepGotFocus()
StepLostFocus()
StopClick(cancel As Integer)
StopCompleted(errcode As Long)
StopGotFocus()
StopLostFocus()

OLE

Properties

Action	Integer
AppIsRunning	Boolean
AutoActivate	Integer
AutoVerbMenu	Boolean
BackStyle	Integer
Class	String
Data	Long
DataChanged	Boolean
DataField	String
DataText	String
DisplayType	Integer
FileNumber	Integer
Format	String
HostName	String
LpOleObject	Long
MiscFlags	Integer
ObjectAcceptFormats	String Array
ObjectGetFormats	String Array
ObjectVerbFlags	Long Array
ObjectVerbs	String Array
ObjectVerbsCount	Integer
OLEDropAllowed	Boolean
OLEType	Integer
OLETypeAllowed	Integer
PasteOK	Boolean
Picture	IPictureDisp Object
SizeMode	Integer
SourceDoc	String
SourceItem	String
UpdateOptions	Integer
Verb	Integer

Methods

Close	object.Close()
Copy	object.Copy()
CreateEmbed	object.CreateEmbed(sourcedoc, [class])
CreateLink	object.CreateLink(sourcedoc, [sourceitem])
Delete	object.Delete()
DoVerb	object.DoVerb([verb])
FetchVerbs	object.FetchVerbs()
InsertObjDlg	object.InsertObjDlg()
Paste	object.Paste()
PasteSpecialDlg	object.PasteSpecialDlg()
ReadFromFile	object.ReadFromFile(filenum)
SaveToFile	object.SaveToFile(filenum)
SaveToOle1File	object.SaveToOle1File(filenum)
Update	object.Update()

Events

ObjectMove(left As Single, top As Single, width As Single, height As Single)
Resize(newheight As Single, newwidth As Single)
Updated(code As Integer)

OptionButton
Properties
Alignment	Integer
Caption	String
DisabledPicture	IPictureDisp Object
DownPicture	IPictureDisp Object
MaskColor	Long
Picture	IPictureDisp Object
Style	Integer
UseMaskColor	Boolean
Value	Boolean

PictureBox
Properties
Align	Integer
AutoRedraw	Boolean
AutoSize	Boolean
ClipControls	Boolean
CurrentX	Single
CurrentY	Single
DrawMode	Integer
DrawStyle	Integer
DrawWidth	Integer
FillColor	Long
FillStyle	Integer
Image	IPictureDisp Object
Picture	IPictureDisp Object

Methods
Circle	object.Circle(step, x, y, radius, color, start, end, aspect)
Cls	object.Cls
Line	object.Line(flags, x1, y1, x2, y2, color)
PaintPicture	object.PaintPicture(picture, x1, y1, [width1], [height1], [x2], [y2], [width2], [height2], [opcode])
Point	object.Point(x, y) As Long
PSet	object.PSet(step, x, y, color)
TextHeight	object.TextHeight(str) As Single
TextWidth	object.TextWidth(str) As Single

Events
Change()
Paint()
Resize()

PictureClip
Properties
CellHeight	Integer
CellWidth	Integer
Clip	IPictureDisp Object
ClipHeight	Integer
ClipWidth	Integer
ClipX	Integer
ClipY	Integer
Cols	Integer
GraphicCell	IPictureDisp Object Array
Picture	IPictureDisp Object
Rows	Integer
StretchX	Integer
StretchY	Integer

Printer
Properties
ColorMode	Integer
Copies	Integer
CurrentX	Single
CurrentY	Single
DeviceName	String
DrawMode	Integer
DrawStyle	Integer
DrawWidth	Integer
DriverName	String
Duplex	Integer
FillColor	Long
FillStyle	Integer
FontCount	Integer
Fonts	String Array
Orientation	Integer
Page	Integer
PaperBin	Integer
PaperSize	Integer
Port	String
PrintQuality	Integer
TrackDefault	Boolean
TwipsPerPixelX	Single
TwipsPerPixelY	Single
Zoom	Long

Methods
Circle	object.Circle(step, x, y, radius, color, start, end, aspect)
EndDoc	object.EndDoc
KillDoc	object.KillDoc
Line	object.Line(flags, x1, y1, x2, y2, color)
NewPage	object.NewPage
PaintPicture	object.PaintPicture(picture, x1, y1, [width1], [height1], [x2], [y2], [width2], [height2], [opcode])
PSet	object.PSet(step, x, y, color)
TextHeight	object.TextHeight(str) As Single
TextWidth	object.TextWidth(str) As Single

ProgressBar

Properties

Align	Integer
Max	Single
Min	Single
Value	Single

Property Page

Properties

ActiveControl	Control Object
AutoRedraw	Boolean
Caption	String
Changed	Boolean
ClipControls	Boolean
Controls	Object Collection
Count	Integer
CurrentX	Single
CurrentY	Single
DrawMode	Integer
DrawStyle	Integer
DrawWidth	Integer
FillColor	Long
FillStyle	Integer
Image	IpictureDisp Object
KeyPreview	Boolean
Palette	IPictureDisp Object
PaletteMode	Integer
Picture	IpictureDisp Object
SelectedControls	SelectedControls Collection

Methods

Circle	object.Circle(step, x, y, radius, color, start, end, aspect)
Cls	object.Cls
Line	object.Line(flags, x1, y1, x2, y2, color)
PaintPicture	object.PaintPicture(picture, x1, y1, [width1], [height1], [x2], [y2], [width2], [height2], [opcode])
Point	object.Point(x, y) As Long
PopupMenu	object.PopupMenu(menu, [flags], [x], [y], [defaultmenu])
PSet	object.PSet(step, x, y, color)
TextHeight	object.TextHeight(str) As Single
TextWidth	object.TextWidth(str) As Single

Events

ApplyChanges()
EditProperty(propertyname As String)
Initialize()
Paint()
SelectionChanged()
Terminate()

RichTextBox
Properties

AutoVerbMenu	Boolean	SelColor	Variant
BulletIndent	Single	SelFontName	Variant
DisableNoScroll	Boolean	SelFontSize	Variant
FileName	String	SelHangingIndent	Variant
HideSelection	Boolean	SelIndent	Variant
Locked	Boolean	SelItalic	Variant
MaxLength	Long	SelLength	Long
MultiLine	Boolean	SelProtected	Variant
OLEObjects	OLEObjects Collection	SelRightIndent	Variant
RightMargin	Single	SelRTF	String
ScrollBars	ScrollBarsConstants Enum	SelStart	Long
		SelStrikethru	Variant
SelAlignment	Variant	SelTabCount	Variant
SelBold	Variant	SelTabs	Variant Array
SelBullet	Variant	SelText	String
SelCharOffset	Variant	SelUnderline	Variant
		Text	String
		TextRTF	String

Methods

Find	object.Find(string, [start], [end], [options])
GetLineFromChar	object.GetLineFromChar(charpos)
LoadFile	object.LoadFile(pathname, [flags])
SaveFile	object.SaveFile(pathname, [flags])
SelPrint	object.SelPrint(hdc)
Span	object.Span(characterset, [forward], [negate])
UpTo	object.UpTo(characterset, [forward], [negate])

Events
Change()
SelChange()

Screen
Properties
ActiveControl	Control Object
ActiveForm	Form Object
FontCount	Integer
Fonts	String Array
TwipsPerPixelX	Single
TwipsPerPixelY	Single

Shape
Properties
BackStyle	Integer
BorderColor	Long
BorderWidth	Integer
DrawMode	Integer
FillColor	Long
FillStyle	Integer
Shape	Integer

Slider
Properties
GetNumTicks	Long
LargeChange	Long
Max	Long
Min	Long
Orientation	OrientationConstants Enum
SelectRange	Boolean
SelLength	Long
SelStart	Long
SmallChange	Long
TickFrequency	Long
TickStyle	TickStyleConstants Enum
Value	Long

Methods
ClearSel	object.ClearSel

Events
Change()
Scroll()

StatusBar
Properties
Align	Integer
Panels	Panels Collection
ShowTips	Boolean
SimpleText	String
Style	SbarStyleConstants Enum

Events
PanelClick(panel As Panel)
PanelDblClick(panel As Panel)

Sysinfo
Properties
ACStatus	Integer
BatteryFullTime	Long
BatteryLifePercent	Integer
BatteryLifeTime	Long
BatteryStatus	Integer
OSBuild	Integer
OSPlatform	Integer
OSVersion	Single
ScrollBarSize	Single
WorkAreaHeight	Single
WorkAreaLeft	Single
WorkAreaTop	Single
WorkAreaWidth	Single

Events
ConfigChangeCancelled()
ConfigChanged(oldconfignum As Long, newconfignum As Long)
DeviceArrival(devicetype As Long, deviceid As Long, devicename As String, devicedata As Long)
DeviceOtherEvent(devicetype As Long, eventname As String, datapointer As Long)
DeviceQueryRemove(devicetype As Long, deviceid As Long, devicename As String, devicedata As Long, cancel As Boolean)
DeviceQueryRemoveFailed(devicetype As Long, deviceid As Long, devicename As String, devicedata As Long)
DeviceRemoveComplete(devicetype As Long, deviceid As Long, devicename As String, devicedata As Long)
DeviceRemovePending(devicetype As Long, deviceid As Long, devicename As String, devicedata As Long)
DevModeChanged()
DisplayChanged()
PowerQuerySuspend(cancel As Boolean)
PowerResume()
PowerStatusChanged()
PowerSuspend()
QueryChangeConfig(cancel As Boolean)
SettingChanged(item As Integer)
SysColorsChanged()
TimeChanged()

TabStrip

Properties

ClientHeight	Single
ClientLeft	Single
ClientTop	Single
ClientWidth	Single
ImageList	Object
MultiRow	Boolean
SelectedItem	Tab Object
ShowTips	Boolean
Style	TabStyleConstants Enum
TabFixedHeight	Integer
TabFixedWidth	Integer
Tabs	Tabs Collection
TabWidthStyle	TabWidthStyleConstants Enum

Events
BeforeClick(cancel As Integer)

TextBox
Properties

Alignment	Integer
HideSelection	Boolean
Locked	Boolean
MaxLength	Long
MultiLine	Boolean
PasswordChar	String
ScrollBars	Integer
SelLength	Long
SelStart	Long
SelText	String
Text	String

Events
Change()

Timer
Properties

Interval	Long

Events
Timer()

Toolbar
Properties

Align	Integer
AllowCustomize	Boolean
ButtonHeight	Single
Buttons	Buttons Collection
ButtonWidth	Single
Controls	Controls Collection
HelpFile	String
ImageList	Object
ShowTips	Boolean
Wrappable	Boolean

Methods

Customize	object.Customize
RestoreToolbar	object.RestoreToolbar(key, subkey, value)
SaveToolbar	object.SaveToolbar (key, subkey, value)

Events
Change()

TreeView
Properties

DropHighlight	Node Object
HideSelection	Boolean
ImageList	Object
Indentation	Single
LabelEdit	LabelEditConstants Enum
LineStyle	TreeLineStyleConstants Enum
Nodes	Nodes Collection
PathSeparator	String
SelectedItem	Node Object
Sorted	Boolean
Style	TreeStyleConstants Enum

Methods

HitTest	object.HitTest(x, y) As Node
StartLabelEdit	object.StartLabelEdit

Events
AfterLabelEdit(cancel As Integer, newstring As String)
BeforeLabelEdit(cancel As Integer)
Expand(node As Node)
NodeClick(node As Node)

UpDown
Properties
Alignment	AlignmentConstants Enum
AutoBuddy	Boolean
BuddyControl	Variant
BuddyProperty	Variant
Increment	Long
Max	Long
Min	Long
OLEDropMode	OLEDropConstants Enum
Orientation	OrientationConstants Enum
SyncBuddy	Boolean
Value	Long
Wrap	Boolean

Events
Change()
DownClick()
UpClick()

Appendix K

ABOUT THE CD-ROM

The CD-ROM included with Learning Visual Basic with Applications includes all of the necessary tools (with the exception of Visual Basic) to write the programs that are developed in each chapter. It also includes full color images of all the figures in the book, and the source code and executable files for the sample projects.

CD FOLDERS:

- **FIGURES**: The full color version of all the figures in the book.
- **SOURCE**: Arranged by chapter and includes the source code and executable files for every sample in the book.
- **PROGRAMS**:
  - The first program available on the CD-ROM is the Microsoft DirectX 8 Software Developers Kit. Located in the DirectX 8 SDK sub-directory, it is the full version of the SDK and is needed for the DirectX model viewer program developed in the book. Once installed, it includes samples and documentation.
  - Microsoft Agent and character files are in the Agent sub-directory. It contains the 4 standard Agent characters Genie, Peedy, Robbie and Merlin and the complete Agent SDK.

SYSTEM REQUIREMENTS:

Microsoft Visual Basic version 5 or above
Windows 95, 98, NT, 2000
Pentium
CD-ROM
Hard Drive: 200 MB of free space to install the Microsoft DirectX SDK and Agent SDK
32 MB of RAM

INSTALLATION:

To use the programs on the CD-ROM, your system should match at least the minimum system requirements. You should contact Microsoft directly if you need any help installing Agent or DirectX. The image files that are used in the development of projects are in JPEG format while the chapter figures are TIFF files.

LICENSE

Microsoft DirectX 8.0 Software Development Kit

This program was reproduced by Charles River Media under a special arrangement with Microsoft Corporation. For this reason, Charles River Media is responsible for the product warranty and for support. If your diskette is defective, please return it to Charles River Media, which wil arrange for its replacement. PLEASE DO NOT RETURN IT TO MICROSOFT CORPORATION. Any product support will be provided, if at all, by Charles River Media. PLEASE DO NOT CONTACT MICROSOFT CORPORATION FOR PRODUCT SUPPORT. End users of this Microsoft program shall not be considered "registered owners" of a Microsoft product and therefore shall not be eligible for upgrades, promotions or other benefits available to "registered owners" of Microsoft products.

END-USER LICENSE AGREEMENT FOR MICROSOFT SOFTWARE

IMPORTANT-READ CAREFULLY: This End-User License Agreement ("EULA") is a legal agreement between you (either an individual or a single entity) and Microsoft Corporation ("Microsoft") for the Microsoft software product identified above, which includes computer software and may include associated media and printed materials, and "online" or electronic documentation ("SOFTWARE PRODUCT"). The SOFTWARE PRODUCT provided to you by Microsoft. Any software provided along with the SOFTWARE PRODUCT that is associated with a separate end-user license agreement is licensed to you under the terms of that license agreement. You agree to be bound by the terms of this EULA by installing, copying, downloading, accessing or otherwise using the SOFTWARE PRODUCT. If you do not agree to the terms of this EULA, do not install or use the SOFTWARE PRODUCT.

SOFTWARE PRODUCT LICENSE

The SOFTWARE PRODUCT is protected by copyright laws and international copyright treaties, as well as other intellectual property laws and treaties. The SOFTWARE PRODUCT is licensed, not sold.

1. GRANT OF LICENSE.

Microsoft grants you the following rights provided that you comply with all the terms and conditions of this EULA:

SOFTWARE PRODUCT. You may install and use the SOFTWARE PRODUCT on up to ten (10) computers, including workstations, terminals or other digital electronic devices ("COMPUTERS"), provided that you are the only individual using the SOFTWARE PRODUCT on each COMPUTER, to design, develop, and test software application products for use with Microsoft operating system products including Windows 2000, Windows 95, Windows 98 and Windows Me and subsequent releases thereto ("Application"). If you are an entity, Microsoft grants you the right to designate one individual within your organization to have the right to use the SOFTWARE PRODUCT in the manner provided above.

SAMPLE CODE. You may modify the sample source code located in the SOFTWARE PRODUCT's <Release Image> root directory "<Release Image>\DXF\Samples\Multimedia"("Sample Code") to design, develop and test your Application. You may also reproduce and distribute the Sample Code in object code form along with any modifications you make to the Sample Code, provided that you comply with the Distribution Requirements described below. For purposes of this Section, "modifications" shall mean enhancements to the functionality of the Sample Code.

REDISTRIBUTABLE CODE. Portions of the SOFTWARE PRODUCT are designated as "Redistributable Code". If you choose to distribute the Redistributable Code, you must include the files as specified listed in the SOFTWARE PRODUCTS <Release Image> root directory "<Release Image>\DXF\doc\directxeulas\directx redist.txt". No other modifications, additions, or deletions to the Redistributable Code are permitted without written permission from Microsoft Corporation. Your rights to distribute the Redistributable Code are subject to the Distribution Requirements described below.

DISTRIBUTION REQUIREMENTS. You may reproduce and distribute an unlimited number of copies of the Sample Code and/or Redistributable Code, (collectively "REDISTRIBUTABLE COMPONENTS")as described above, provided that (a) you distribute the REDISTRIBUTABLE COMPONENTS only as part of, or for use in conjunction with your Application; (b) your Application adds significant and primary functionality to the REDISTRIBUTABLE COMPONENTS; (c) the REDISTRIBUTABLE COMPONENTS only operate in conjunction with Microsoft Windows operating system products including Windows 2000, Windows 95, Windows 98, Windows Me, and subsequent versions thereof, (d) you distribute your Application containing the REDISTRIBUTABLE COMPONENTS pursuant to an End-User License Agreement (which may be "break-the-seal", "click-wrap", or signed), with terms no less protective than those contained herein; (e) you do not permit further redistribution of the REDISTRIBUTABLE COMPONENTS by your end-user customers; (f) you must use the setup utility included with the REDISTRIBUTABLE COMPONENTS to install the Redistributable Code; (g) you do not use Microsoft's name, logo, or trademarks to market your Application; (h) you include all copyright and trademark notices contained in the REDISTRIBUTABLE COMPONENTS; (i) you include a valid copyright notice on your Application; and (j) you agree to indemnify, hold harmless, and defend Microsoft from any against any claims or lawsuits, including attorneys' feeds, that arise or result from the use or distribution of your Application.

If you distribute the Redistributable Code separately for use with your Application (such as on your web site or as part of an update to your Application), you must include an end user license agreement in the install program for the Redistributable Code in the form of <Release Image>\license\directx end user eula.txt. Contact Microsoft for the applicable royalties due and other licensing terms for all other uses and/or distribution of the REDISTRIBUTABLE COMPONENTS.

2. COPYRIGHT.

All title and intellectual property rights in and to the SOFTWARE PRODUCT (including but not limited to any images, photographs, animations, video, audio, music, text and "applets," incorporated into the SOFTWARE PRODUCT), any accompanying printed materials, and any copies of the SOFTWARE PRODUCT, are owned by Microsoft or its suppliers. All title and intellectual property rights in and to the content which may by accessed through use of the SOFTWARE PRODUCT is the property of the respective content owner and may be protected by applicable copyright or other intellectual property laws and treaties . This EULA grants you no rights to use such content. If this SOFTWARE PRODUCT contains documentation which is provided only in electronic form, you may print one copy of such electronic documentation. You may not copy the printed materials accompanying the SOFTWARE PRODUCT. All rights not expressly granted are reserved by Microsoft.

3. DESCRIPTION OF OTHER RIGHTS AND LIMITATIONS.

   a. Limitations on Reverse Engineering, Decompilation and Disassembly. You may not reverse engineer, decompile, or disassemble the SOFTWARE PRODUCT, except and only to the extent that such activity is expressly permitted by applicable law notwithstanding this limitation.
   b. Rental. You may not rent, lease or lend the SOFTWARE PRODUCT.

   c. Support Services. Microsoft may provide you with support services related to the SOFTWARE PRODUCT ("Support Services"). Use of the Support Services is governed by the Microsoft policies and programs described in the user manual, in "on line" documentation and/or other Microsoft-provided materials. Any supplemental software code provided to you as part of the Support Services shall be considered part of the SOFTWARE PRODUCT and subject to the terms and conditions of this EULA. With respect to technical information you provide to Microsoft as part of the Support Services, Microsoft may use such information for its business purposes, including for product support and development. Microsoft will not utilize such technical information in a form that personally identifies you.

   d. Software Transfer. The initial user of the SOFTWARE PRODUCT may make a one-time permanent transfer of this EULA and SOFTWARE PRODUCT only directly to an end-user. This transfer must include all of the SOFTWARE PRODUCT (including all component parts, the media and printed materials, any upgrades, this EULA, and, if applicable, the Certificate of Authenticity). Such transfer may not be by way of consignment or any other indirect transfer. The transferee of such one-time transfer must agree to comply with the terms of this EULA, including the obligation not to further transfer this EULA and SOFTWARE PRODUCT.

   e. Termination. Without prejudice to any other rights, Microsoft may cancel this EULA if you do not abide with the terms and conditions of this EULA, in which case, you must cease all use or distribution and destroy all copies of the SOFTWARE PRODUCT and all of its component parts.

4. U.S. GOVERNMENT RESTRICTED RIGHTS.

All SOFTWARE PRODUCT provided to the U.S. Government pursuant to solicitations issued on or after December 1, 1995 is provided with the commercial license rights and restrictions described elsewhere herein. All SOFTWARE PRODUCT provided to the U.S. Government pursuant to solicitations issued prior to December 1, 1995 is provided with "Restricted Rights" as provided for in FAR, 48 CFR 52.227-14 (JUNE 1987) or DFAR, 48 CFR 252.227-7013 (OCT 1988), as applicable. The reseller is responsible for ensuring SOFTWARE PRODUCT is marked with the "Restricted Rights Notice" or "Restricted Rights Legend", as required. All rights not expressly granted are reserved.

5. EXPORT RESTRICTIONS.

You acknowledge that the SOFTWARE PRODUCT is of U.S.-origin. You agree to comply with all applicable international and national laws that apply to these products, including the U.S. Export Administration Regulations, as well as end-user, end-use and country destination restrictions issued by U.S. and other governments. For additional information see http://www.microsoft.com/exporting/.

6. DISCLAIMER OF WARRANTIES.

TO THE MAXIMUM EXTENT PERMITTED BY APPLICABLE LAW, MICROSOFT AND ITS SUPPLIERS PROVIDE TO YOU THE SOFTWARE PRODUCT AND SUPPORT SERVICES (IF ANY) AS IS AND WITH ALL FAULTS; AND HEREBY DISCLAIM ALL OTHER WARRANTIES AND CONDITIONS, EITHER EXPRESS, IMPLIED OR STATUTORY, INCLUDING, BUT NOT LIMITED TO, ANY (IF ANY) WARRANTIES OR CONDITIONS OF MERCHANTABILITY, OF FITNESS FOR A PARTICULAR PURPOSE, OF LACK OF VIRUSES, OF ACCURACY OR COMPLETENESS OF RESPONSES, OF RESULTS, AND OF LACK OF NEGLIGENCE OR LACK OF WORKMANLIKE EFFORT, ALL WITH REGARD TO THE SOFTWARE PRODUCT, AND THE PROVISION OF OR FAILURE TO PROVIDE SUPPORT SERVICES. ALSO, THERE IS NO WARRANTY OR CONDITION OF TITLE, QUIET ENJOYMENT, QUIET POSSESSION, AND CORRESPONDENCE TO DESCRIPTION OR NON-INFRINGEMENT WITH REGARD TO THE SOFTWARE PRODUCT.

7. EXCULSION OF INCIDENTAL, CONSEQUENTIAL AND CERTAIN OTHER DAMAGES.

TO THE MAXIMUM EXTENT PERMITTED BY APPLICABLE LAW, IN NO EVENT SHALL MICROSOFT OR ITS SUPPLIERS BE LIABLE FOR ANY SPECIAL, INCIDENTAL, INDIRECT, OR CONSEQUENTIAL DAMAGES WHATSOEVER (INCLUDING, BUT NOT LIMITED TO, DAMAGES FOR LOSS

OF BUSINESS PROFITS OR CONFIDENTIAL OR OTHER INFORMATION, FOR BUSINESS INTERRUPTION, FOR PERSONAL INJURY, FOR LOSS OF PRIVACY, FOR FAILURE TO MEET ANY DUTY (INCLUDING OF GOOD FAITH OR OF REASONABLE CARE), FOR NEGLIGENCE, AND FOR ANY OTHER PECUNIARY OR OTHER LOSS WHATSOEVER) ARISING OUT OF OR IN ANY WAY RELATED TO THE USE OF OR INABILITY TO USE THE SOFTWARE PRODUCT, THE PROVISION OF OR FAILURE TO PROVIDE SUPPORT SERVICES, OR OTHERWISE UNDER OR IN CONNECTION WITH ANY PROVISION OF THIS EULA, EVEN IN THE EVENT OF THE FAULT, TORT (INCLUDING NEGLIGENCE), STRICT LIABILITY, BREACH OF CONTRACT OR BREACH OF WARRANTY OF MICROSOFT OR ANY SUPPLIER, AND EVEN IF MICROSOFT OR ANY SUPPLIER HAS BEEN ADVISED OF THE POSSIBILITY OF SUCH DAMAGES.

8. LIMITATION OF LIABILITY AND REMEDIES.

NOTWITHSTANDING ANY DAMAGES THAT YOU MIGHT INCUR FOR ANY REASON WHATSOEVER (INCLUDING, WITHOUT LIMITATION, ALL DAMAGES REFERENCED ABOVE AND ALL DIRECT OR GENERAL DAMAGES), THE ENTIRE LIABILITY OF MICROSOFT AND ANY OF ITS SUPPLIERS UNDER ANY PROVISION OF THIS EULA AND YOUR EXCLUSIVE REMEDY FOR ALL OF THE FOREGOING SHALL BE LIMITED TO THE GREATER OF THE AMOUNT ACTUALLY PAID BY YOU FOR THE SOFTWARE PRODUCT OR U.S.$5.00. THE FOREGOING LIMITATIONS, EXCLUSIONS AND DISCLAIMERS DESCRIBED ABOVE SHALL APPLY TO THE MAXIMUM EXTENT PERMITTED BY APPLICABLE LAW, EVEN IF ANY REMEDY FAILS ITS ESSENTIAL PURPOSE.

9. APPLICABLE LAW.

If you acquired this SOFTWARE PRODCUT in the United States, this EULA is governed by the laws of the State of Washington.

If you acquired this SOFTWARE PRODUCT in Canada, this EULA is governed by the laws of the Province of Ontario, Canada. Each of the parties hereto irrevocably attorns to the jurisdiction of the courts of the Province of Ontario and further agrees to commence any litigation which may arise hereunder in the courts located in the Judicial District of York, Province of Ontario.

If this product was acquired outside the United States, then local law may apply.

10. QUESTIONS?

Should you have any questions concerning this Agreement, or if you desire to contact Microsoft for any reason, please contact the Microsoft subsidiary serving your country, or write: Microsoft Sales Information Center/One Microsoft Way/Redmond, WA 98052-6399.

Index

Accept method, 172
Accept requestID, 176
ActiveX
  code to close previous Winsock connections, 197
  code to set local port equal to 0, 197
  connecting to an Internet server, 197
  retrieving date and time from Internet server, 197
ActiveX control properties, 199
ActiveX controls, 8
  adding new properties, 199
  assigning value with Let property, 199
  compiling, 200
  creating a new project utilizing control, 200
  custom controls, 195
  properties so end user can modify values, 199
  properties so end user can receive information from the control, 199
  receiving values with Get property, 199
  testing, 199–201
adding a form to a project, 10

adding Agent control to a project, 183
adding button value to named Text Box, 252
adding Common Dialog Control to toolbox, 41
adding controls to the toolbox, 23
adding forms to a project, 160
adding in VB calculator, 254
adding INET control to toolbox, 144
adding new form to project, 173
adding Rich Textbox control to toolbox, 128
AddItem method, 46
AddItem method for Drawing Styles Combo Box, 208
address, 62
Agent
  Caption property parameter, 188
  command syntax, 188
  Confidence value, 190
  loading Agent Character file, 184
  Name parameter, 188

  parameters for adding commands, 187
  responding to speech output, 187
  SCROLL LOCK key, 187
  Select Case for commands, 192
  Shell command for executing Agent commands, 192
  Speak method, 186
  TTS, 186
  Voice property parameter, 188
Agent chore, 182
Agent command UserInput object, 190
Agent variables, 184
Agent web site, 182
Agent1.Characters.Load command, 184
aligning in code, 137
animated Desktop Assistant, 181
animation tool for querying individual Agent, 186
API
  defined, 60
  initializing the window regions, 64

435

reason to use, 61
storage, 60
API calls, 44
API cautions, 60
API declarations, 135
API declarations and variables for Paint program, 207
API function for pasting text and pictures, 135
API Text Viewer, 61
arranging windows code, 163
array
 defined, 44
 index number, 44
array of Command Buttons, 244
array of Label controls, 262
array of points to draw arrow shape, 68
arrow shaped buttons, 68
Asc function, 49
ASCII files, 128
assigning Genres types to combo box, 46
authentication, 144
BackColor property, 65
BackStyle property, 67
binary reading, 49
Bind method, 172
bit block transfers, 235, 237
BitBlt, 237
BitBlt API variables, 236
BitBlt functions, 237
BitBlt variables, 237
Boolean variable, 50
Border Style property, 66
borders for Pong game, 82
BYTE bScan, 234

BYTE bVk virtual-key code, 234
BytesReceived property, 172
Caesar Cipher encryption, 272
calling a DLL procedure, 61
Camera, 100
canceling the INET command, 151
Caption property, 24
Caption property parameter for Agent, 188
Case statement for Agent commands, 192
cdlCFBoth, 136
cdlCFPrinterFonts, 136
cdlCFScreenFonts, 136
CDUP command, 150
change size of drawing window, 208
change width of tool you are using, 208
changing appearance of buttons and form, 69
changing background color of Rich Textbox in code, 136
changing colors in Paint program, 214
changing drawing styles dynamically, 208
Chat program, 171
Check Box control, 7
CheckWin procedure, 267
child forms
 adding to a project, 160
 captions, 160
 dynamic creation at runtime, 161
child windows, 158

clearing Text Boxes, 256
clearing the Clipboard, 232
clearing the screen, 237
Click procedure, 17, 29
client form, 173
 code for Connect button, 177
 code for receiving data from server, 177
 code for Send button, 177
 code for sending data to server, 177
 DataArrival, 177
 RemoteHost property, 177
 RemotePort property, 177
Clipboard
 code for screen capture, 232
 code to clear, 232
Close method, 172
Close statement, 49
closing a form, 71
closing a program, 19
closing the default form, 159
code for alignment, 137
code for changing Background color of Rich Textbox, 136
code for checking for data before printing, 134
code for control array, 251
code for displaying pop-up menu with right-click, 215
code for displaying printer and screen fonts, 136
code for printing without page numbers, 134
code to arrange windows, 163
code to open files, 133

Index

code to retrieve time from new user control, 201
code to save files in RTF, 134
code to set default property of printer, 134
code upgrading, 12
Code window, 9
  opening, 9
color selector in Paint program, 210
combining regions, 64
combo box
  Add Item method, 46
Combo Box control, 7
Command Button control, 7
command line arguments, 122
commands for assigning drawing tool, 213
comments
using temporary comments to test an application, 121
Common Dialog Control
  adding to toolbox, 41
  initializing, 46
  purpose, 41
Common Dialog Control Flags property, 136
Common Dialog filter, 133
  limiting files to text and richtext, 133
comparing time with future time, 266
compiling an ActiveX control, 200
compiling and renaming a screen saver, 123
Components window, 23

Confidence value for Agent command, 190
Connect method, 172
connection request, 176
constant for Paste command, 135
context menus, 10
control array, 246, 262
  index value, 265
  variable declaration, 264
control array code, 251
converting text to speech, 186
converting the Degree values to radians, 256
coordinate system for Direct3D, 99
copying and pasting buttons, 245
copying and pasting labels, 262
Cos, 253
counters for mouse movement, 121
counting characters and converting with XOR, 275
Create an ActiveX Control option, 196
creating multiple forms in a project, 160
creating new files with code, 132
creating shapes, 216
custom ActiveX control, 195
custom toolbar, 6
custom toolbars, 205–6
customizing built-in toolbars, 6
Data control, 8
DataArrival event, 198

Debug toolbar, 5
decimal points in VB calculator, 254
decimal removal, 118
Declare statement, 61
Declare statements keywords, 61
declaring Agent control reference at runtime, 184–85
declaring DLLs, 61
declaring variables, 18
declaring variables for Agent, 184
default startup object, 35
default TTS information, 184
delaying a program, 267
designing your own Agents, 182
determining color in Paint program, 211
Dim statement, 18
Direct 3D
  object rotation, 100
Direct3D
  advantages, 98
  camera, 100
  closing the application, 107
  coordinate system, 99
  defined, 98
  displaying information for user during rendering, 106
  Immediate Mode, 100
  initializing, 104
  light color, 105
  light position, 105
  lights, 101
  modes, 100
  references, 99
  rendering the scene, 106
  Retained Mode
    declarations, 102

scenes, 100
setting quality of render, 104
stopping main rendering loop, 102
tracking number of displayed frames, 102
using hardware card or Software Renderer, 104
viewport resolution, 105
DirectDraw
  declarations, 102
  initializing, 103
  preparing for full screen display, 103
  setting surfaces, 102
  setting up surfaces, 103
Directory List Box control, 8
DirectX
  declarations, 101
  defined, 98
  form load, 102
  initializing, 102
  opening the .X file, 105
  positioning .X file in center of scene before rotating, 106
  references, 99
  Retained Mode, 100
disabling Spin button until previous spin completed, 265
displaying multiple child windows in the MDI parent, 162
displaying printer and screen fonts through code, 136
dividers in menus, 130
dividing in VB calculator, 254
DLLs
  defined, 60
  function syntax, 61
  procedure syntax, 61

reason to use, 61
storage, 60
Document, 144
DoEvents, 234
DoEvents command, 266
doubles in VB calculator, 253
downloading Agents, 182
drawing gradients, 119
drawing to exact points on the Picture Box, 214
drawing to the Picture Box, 214–18
drawing tool command code, 213
drawing tool style, 211
Drive List Box control, 8
DWORD dwExtraInfo, 234
DWORD dwFlags, 234
dynamic creation of child forms, 161
dynamically changing Web Browser form caption, 167
Edit Toolbar, 5
Ellipse, 207
empty strings, 49
Enabled property, 86
encryption
  Caesar Cipher, 272
  counting characters and converting with XOR, 275
  creating random encryption key, 275
  Randomize function, 275
  returning to original value, 273
  XOR, 276
  XOR Boolean operator, 272
encryption purpose, 271
encryption variables, 274

End Sub line, 18
error handling, 198
Error Resume Next statement, 184
ErrorHandler, 47
event, 16
event procedure, 28
events, 28
Execute command, 147
executing a program, 18, 35
Exit command code, 161
exiting subroutine if user clicks Cancel, 134
File List Box control, 8
File Transfer Protocol, 143
File|Add New Project, 200
FileName property, 31
filling in Picture Box, 214
font
  resetting to Normal through code, 137
For...Next Loop, 46
ForeColor property, 120
form arrange integers, 163
form background color for Pong game, 82
form caption
  setting in code, 133
Form Designer window, 6, 9
Form Editor Toolbar, 5
Form Layout window, 6, 11
Form Load, 28
Form Load event, 17, 18
Form Load vs. Form Activate, 118
Form Unload, 34, 52
Form window, 6, 9
Form_Click event, 120
Form_KeyDown, 86
Form_KeyDown event, 120
Form_KeyUp, 87

Form_MouseMove event, 120
forms
  purpose, 9
Frame control, 7
frames, 24
frames and option buttons, 27
FTP
  address of computer, 144
  authentication, 144
  basics, 144
  canceling and ending program, 151
  canceling the INET command, 151
  CDUP command, 150
  changing directories, 144
  connecting to server, 144
  disconnecting, 151
  downloading files, 144, 151
  file retrieval, 151
  Form Unload event, 151
  listing of a directory, 150
  password, 144
  privileges, 144, 148
  retrieving data, 150
  retrieving list of files, 144
  user name, 144
FTP program
  connecting to server, 147
  GUI, 145
  variables, 146
Future as Date value, 266
GDI32.DLL, 60
General Declarations, 61
generating random colors, 118
Genie character, 184
Genie.Commands.Add method, 188
Get property, 199
Get statement, 49

GetActiveWindow, 237
GetData command, 149
GetData method, 172
GetData method for client, 177
GetData method for retrieving graphics from Clipboard, 135
GetData method for server, 176
GetDesktopWindow, 237
GetScreen procedure, 237
GetText method for retrieving text from Clipboard, 135
GetWindowDC, 237
GetWindowRect, 237
GoBack method, 164
GoForward method, 164
GoHome method, 164
GotFocus, 252
Gourad Render Mode, 101
GradientForm procedure, 119
Graphical User Interface, 2
GUI, 2
handle, 62
hard-coded file vs. Open dialog box, 31
height of drawing canvas, 211
hiding pop-up menu, 208
how code works, 18
ID3 tags, 48
IDE
  windows, 6–11
IDE components, 2–11
Image control, 8
Immediate Mode, 100
index number, 44

index value, 244, 251, 252
index value of label array, 265
INET, 143
INET command
  canceling, 151
INET commands, 147
INET control
  Document, 144
  monitoring with StateChanged event, 148
  Password, 144
  properties, 144
  RemotePort, 144
  URL, 144
  UserName, 144
INET GetChunk method, 149
InitializeDirectX, 103
InitializeScene, 103
  create camera, 105
  create viewport, 105
  set frames, 105
initializing controls, 29
initializing Randomize function, 118
initializing the Common Dialog Control, 46
initializing the scene, 102
Int function, 118
Integrated Development Environment, 2–11
Interval property for timer control in Pong game, 84
IP address, 172
KERNEL32.DLL, 60
key value for encryption, 272
Keybd Event parameters, 234
KeybdEvent API variables, 234

keyboard shortcuts in
   menus, 130
keywords, 61
Label control, 7
label control array
   assigning random numbers, 264
   setting random value, 265
labeling text box contents, 43
Left property, 65
Let property, 199
lights, 101
Line control, 8, 264
List Box control, 7
Listen method, 172
Listen property of server, 176
loading Agent Character file, 184
LocalHost property, 172
LocalIP property, 172
LocalPort property, 172
LocalPort property of server, 176
logical operators for
   combining regions, 64
looping animation, 187
making objects
   transparent, 67
MCI commands, 40
mciSendString function, 44
   to open file, 46
MDI, 3, 157
   adding the parent form, 159
   child form menus, 160
   code to arrange windows, 163
   creating child forms
      dynamically at
      runtime, 160

dynamic creation of child
   forms, 161
parent form menus, 160
setting the MDIChild
   property, 159
MDI vs. SDI, 158
MDIForm1.Arrange method, 163
menu bar, 4
Menu Editor
   Caption property, 129
   caption using & character, 130
   dividers, 130
   menu item levels, 130
   Name property, 129
   opening, 129
   shortcut key combinations, 130
menu events code, 209
menus
   Alt key access, 130
   command placement, 129
   placement on menu bar, 129
   purpose, 129
message box format, 33
message box in Query
   Unload, 32
message boxes, 18
Microsoft
standard animation files, 186
Microsoft Agent
   installing, 182
   playing animation files, 186
   purpose, 182
   tools, 182
Microsoft Agent ActiveX
   control, 183
Microsoft Agent control, 181
Microsoft Agent web site, 182

Microsoft DirectX8
   Software Developers Kit, 98
Microsoft Internet
   Controls, 164
Microsoft Internet Transfer
   Control, 143
Microsoft Multimedia
   Control, 22
Microsoft SAPI, 182
Microsoft Speech
   Recognition Engine, 182
minimizing and
   maximizing child forms, 160
modifying the way words
   are spoken, 186
MouseDown event, 70
MouseDown event to
   determine color, 211
MouseMove event to
   determine color, 211
MouseUp
   drawing lines, rectangles,
      ellipses, or text, 215
MouseUp event to
   determine color, 211
move and size Rich Textbox
   to match form, 138
MoveForm Mouse
   procedure, 70
MoveTo method for Agent
   placement, 189
moving a form without a
   title bar, 70
moving controls, 14
moving Paint toolbars at
   runtime, 212
moving Picture Box, 214
MP3, 39

displaying Tag information in text boxes, 42
ID3 tags, 48
reading Tag information, 47
Tag information, 40
Tag variable, 48
MultiLine property, 173
Multimedia Control Interface commands, 40
Multimedia Control_Status Update event, 30
Multimedia controls, 25
Multiple Document Interface, 3
multiple forms for a project, 160
Multiple-Document Interface, 157
multiplying in VB calculator, 254
Name parameter for Agent, 188
Name property, 26
naming variables, 17
Navigate method, 164
Network Time Protocol, 198
non-looping animation, 187
NTP, 198
numbers in a string, 254
Object Linking and Embedding, 229
Object Linking and Embedding control, 8
object rotation, 100
OCX controls, 8
OLE, 229
OLE control, 8
OLE custom controls, 8
On Error Resume Next, 48
Open command in code, 31

opening Code window for menu items, 132
opening files through code, 133
opening the Code window, 9, 16
opening Visual Basic, 2
Option Button control, 7
option buttons in frames, 27
Paint program
 user-determined width and height for canvas, 211
parent form
 adding to project, 159
 listing the child windows, 161
parent window, 158
password, 144
PeekData method, 173
performing functions according to selected tool value, 214
pi value, 256
Picture Box control, 7
Picture Box control vs. Image control, 8
picture boxes for Pong game, 82
placing controls on a form, 14
Pointer tool, 7
Pong game
 background color, 82
 ball shape, 82
 ball-paddle collisions, 88
 ball-wall collisions, 87
 borders, 82
 Interval property, 84
 keeping focus on form, 86
 Losses variable, 90
 paddle movement, 86, 89

 paddle position and speed, 85
 paddle-wall collisions, 89
 picture boxes, 82
 resetting game after scoring, 90
 scoring, 90
 Select Case for key action, 87
 stopping paddle movement, 87
 timer, 84
 timer for ball movement, 87
 variables and declarations, 85
 win and loss messages, 90
 Wins and Losses labels, 84
 Wins variable, 90
pop-up menu
 code to make invisible on screen, 208
pop-up menu items, 207
pop-up menu items code, 209
pop-up menus, 10
Port 13, 198
positive or negative toggle function, 253
preventing system crash from use of API, 60
printer fonts, 136
printer handle, 134
printing through code, 134
Private keyword, 61
Private Sub line, 17
procedure scope, 61
procedures, 28
Project Explorer, 10
 keyboard shortcut, 10
 tree structure, 10
Project Explorer window, 6
Project Properties

Command Line Arguments, 122
properties
  setting vs. retrieving, 199
Properties window, 6, 11
  keyboard shortcut, 11
protocol, 172
  TCP, 172
  UDP, 172
PSet, 256
Public keyword, 61
Query Unload, 32
random encryption key, 275
random numbers for label array, 264
random numbers syntax, 118
RandomColors subprocedure, 118
Randomize, 118, 267
Randomize function, 275
randomizing control array, 265
reading size and position of a window, 237
reading the Tag information for MP3 files, 47
ReadTag subroutine, 46
Refresh method, 164
regions, 62
  combining, 64
ReleaseCapture, 207
ReleaseDC, 237
releasing device context, 237
RemoteHost property, 172
RemotePort, 144
RemotePort property, 172
removing a form, 159
removing decimals, 118
removing trailing spaces, 50
renaming controls, 26, 43
Render Modes, 101

render quality, 104
rendering stop, 102
RenderLoop subprocedure, 106
resetting font to Normal through code, 137
resetting initial values of program, 256
resizing controls, 15
resizing the Web Browser control, 166
Retained Mode, 100
  application frame rates, 100
  Microsoft support, 100
retrieving date and time from web sites, 195
returning device context of window, 237
returning handle to active window, 237
returning handle to desktop window, 237
Rich Text Format, 128
Rich Textbox ActiveX control, 128
Rnd function, 118
RTF, 128
RTrim function, 50
running a program, 18
running a screen saver, 123
running an Agent application on a computer without Agent support, 183
running VB6 and VB.NET simultaneously, 13
Save Before Run option, 60
saving a program, 19
saving files through code, 134
saving projects, 34

scene initializing, 102
scenes, 100
sckClosed, 176
scope, 61
screen capture
  BitBlt, 235–39
  BitBlt functions, 237
  code for Clipboard, 232
  displaying contents of Clipboard on form, 232
  full screen vs. active window, 230
  GetScreen procedure, 237
  KeybdEvent, 232–35
  KeybdEvent parameters, 234
  Print Screen key, 230–32
  program techniques, 229
screen fonts, 136
screen saver
  adding a command line parameter to IDE, 122
  compiling and renaming, 123
  configuration form, 116
  counters for mouse movement, 121
  drawing gradients, 119
  executing, 123
  foreground color, 120
  form load, 116
  generating random colors, 118
  gradient background, 115
  initializing Randomize function, 118
  label properties, 115
  main form, 114
  moving label at timed intervals, 120
  parameters, 116
  password, 117
  Preview Mode, 117
  problems with running in IDE, 121

Index

RandomColors subprocedure, 118
redrawing the screen at timed intervals, 118
requirements, 113
stopping, 120
text information, 115
timer intervals, 116
variable declaration, 116
Scroll Bar controls, 7
SCROLL LOCK key and Agent, 187
ScrollBars property, 173
SDI, 157
SDI vs. MDI, 158
SDK, 98
  references, 98
Select Case, 87
Select Case for Agent commands, 192
SendData method, 173
SendData method for client, 177
SendData method for server, 176
sending message from client to server, 178
SendMessage, 207
SendMessage API call, 135
server form, 173
  code for receiving data from client, 176
  code for Send button, 176
SetData method for cutting or copying graphics to Clipboard, 135
SetText method for cutting or copying text to Clipboard, 135
setting a filter for opening specific files, 133
setting alignment in code, 137
setting caption of server form dynamically, 176
setting captions, 24
setting default property of printer in code, 134
setting default startup object, 35
setting device type for multimedia player, 31
setting drawing surfaces, 102
setting font attributes in code, 137
setting form caption in code, 133
setting height and width of window, 238
setting the 3D scene, 102
Shape control, 8, 264
shape control for Pong game ball, 82
Shell command for executing Agent commands, 192
Show command
  displaying BitBlt form, 235
  displaying Keybd Event form, 233
  displaying Print Screen form, 232
Show method for Agent character, 184
Show method for client and server forms, 174
ShowColor method, 136
ShowFont method, 136
showing and hiding Tag information, 71
ShowOpen method, 133
ShowPrinter method, 134
ShowSave method, 134
Sin, 253
Single Document Interface, 157
sizing handles, 15
slot machine
  checking for minimum bet, 265
  CheckWin procedure, 267
  delaying program, 265
  message for minimum bet, 265
  program delay, 267
  Randomize function, 267
  temporarily disabling spinner, 265
  wins and losses, 267
slot machine spins, 264–67
SocketHandle property, 172
Software Renderer, 104
Solid Render Mode, 101
Speak method for Agent, 186
Speech API, 182
Speech Control Panel, 182
speech output quality, 186
speech output tags, 186
speech recognition accuracy, 190
Speech Recognition Engine, 187
standard 3D programming interface, 98
standard Microsoft Agent characters, 182
standard toolbar, 4
starting a program, 18
startup form, 174
  displaying server and client forms, 174
  positioning server and client forms, 174
State property, 172

StateChanged event, 148
Stop command, 187
stop files from playing, 52
Stop method, 164
stopping a looping animation, 187
stopping a program, 19
substituting software for hardware features, 98
subtracting in VB calculator, 254
synchronizing dates and times on networks, 198
Tag information showing and hiding, 71
Tag information for MP3 file, 40
Tag variable, 48
Tan, 253
TCP protocol, 172
Testing an application, 34
testing an application using temporary comments, 121
testing the Chat program, 177
Text Box control, 7
Text Boxes
  MultiLine property, 173
  Scroll Bar property, 173
text boxes for MP3 Tag information, 42
Text-To-Speech aspect of Agent, 186
Text-To-Speech engine, 182
third-party controls vs. VB intrinsic controls, 8
Timer control, 8
  ball movement, 87
timer control for Pong game, 84

title bar, 4
toolbars, 4–6
  built-in, 4
  custom, 6
  customizing appearance of built-in toolbars, 6
  Debug toolbar, 5
  Edit toolbar, 5
  Form Editor toolbar, 5
  individual, 5
  moving dynamically at runtime, 212
  standard toolbar, 4
toolbox, 6–8
  adding tools, 23
Toolbox window, 6
Top property, 65
tracking mouse button clicks, 208
tracking mouse position on Picture Box, 208
tracking tool in use, 208
trig functions, 253
trig graphs, 255–56
  drawing line to determine values, 256
  setting scale for output, 256
trim leading or trailing spaces, 198
Trim$ function, 198
TTS, 182
TwipsPerPixelX for form placement, 189
TwipsPerPixelY for form placement, 189
UDP protocol, 172
underscore character in menu names, 130
underscore character in VB, 34
Undo command, 214
updating status of Multimedia control, 30, 32

Upgrade Wizard, 12
upgrading code, 12
URL, 144
user name, 144
USER32.DLL, 60
UserControl form, 196
UserControl_Initialize event, 197
UserInput.Alt1Name property, 191
UserName, 144
user-selectable drawing tool style, 211
user-selectable width of line, 211
user-selected width and height for drawing space, 211
using temporary comments to test an application, 121
Val keyword, 254
variables
  declaring, 18
  defined, 17
  naming, 17
Variant data type, 44
VB Clipboard object, 135
VB.NET changes from VB6, 12
vbCFDIB format, 135
vbCFText format, 135
vbYesNo button argument, 33
viewport resolution, 105
virtual-key constant, 234
Visible property, 29
Visual Basic and 2D games, 81
Visual Basic and 3D games, 81

Visual Basic time functions, 266
Visual Basic windows, 6–11
Voice Commands window, 187
Voice property parameter for Agent, 188
Web Browser control commands, 164
Web Browser control size and placement, 164
Web Browser form caption, 167
Web Browser GUI, 165
Web Browser methods and properties, 166
WebBrowser StatusTextChange, 166
While...Wend loop, 266
width of drawing canvas, 211
width of drawing tool, 211
Width property, 65
Win and Loss labels for Pong game, 84
WIN32API.TXT, 62
WindowList, 161
windows, 6–11
　Code window, 9
　Form Layout window, 11
　Form window, 9
　Project Explorer, 10
　Properties window, 11
Windows Application Programming Interface calls, 44
winmm.dll file, 40
Winsock control, 171–73
　Chat program, 171
　code to determine what to do about connection, 176

Winsock methods, 172–73
Winsock properties, 172
Winsock1_Error event, 198
Wireframe Render Mode, 101
With...End With statement, 147
word processing formats, 128
word processor features, 127
XOR
　defined, 272
　returning to original value, 273
　True value, 272
XOR Boolean operator, 272
XOR encryption, 272–76